International Symposium on New Developments in Methods and Applications of Few-Body Physics 2017

In Memory of Professor SA Sofianos

Journal of Phzsics: Conference Series Volume 915

Pretoria, South Africa
2 - 5 May 2017

Editors:

G.J. Rampho
M.L. Lekala

ISBN: 978-1-5108-5068-2
ISSN: 1742-6588

Printed from e-media with permission by:

Curran Associates, Inc.
57 Morehouse Lane
Red Hook, NY 12571

Some format issues inherent in the e-media version may also appear in this print version.

Printed by Curran Associates, Inc. (2017)

For permission requests, please contact the Institute of Physics
at the address below.

Institute of Physics
Dirac House, Temple Back
Bristol BS1 6BE UK

Phone: 44 1 17 929 7481
Fax: 44 1 17 920 0979

techtracking@iop.org

Additional copies of this publication are available from:

Curran Associates, Inc.
57 Morehouse Lane
Red Hook, NY 12571 USA
Phone: 845-758-0400
Fax: 845-758-2633
Email: curran@proceedings.com
Web: www.proceedings.com

Table of contents

Volume 915

International Symposium on New Developments in Methods and Applications of Few-body Physics: In Memory of Professor SA Sofianos

2–5 May 2017, University of South Africa

Published online: 30 October 2017

Preface

Papers

IOP Conf. Series: Journal of Physics: Conf. Series **915** (2017) 011001 doi:10.1088/1742-6596/915/1/011001

Preface

The symposium was held in memory of Professor S A Sofianos, a distinguished physicist, who dedicated all his academic life to serving the University of South Africa and the Few-body physics community. Fourteen researchers from Brazil, Russia, Japan and South Africa presented their work at the symposium, while many others who were interested could not participate because of financial and visa challenges. Almost all the participants have collaborated with Professor S A Sofianos in their research.

A wide range of few-body physics topics were discussed during the symposium. Nuclear interactions as well as bound and scattering processes formed the bases of all the discussions of the symposium. The long-standing Coulomb problem and the almost forgotten Marchenko inversion resurfaced. New developments in few-body methods such as the Artificial Neural Networks, Integrodifferential Equations Approach, Gaussian Expansion Method and the Faddeev Approach, which are applicable to stable nuclear systems, were discussed. Various three- and four-body nuclear systems were considered for illustrations of these methods. Developments in the Jost function and Jost matrix methods and their application to unstable exotic nuclear systems were also discussed. Professor S A Sofianos worked on all, and many more, of these methods and few-body systems.

The organizers would like to thank the University of South Africa (UNISA) and the National Institute for Theoretical Physics for financial support. We are grateful for the substantial financial contribution from Professor M Braun (UNISA Research Chair: High Performance Scientific Computing) without which the symposium would not have materialized. Logistical support from the Department of Physics, Deanery and Marketing Division of the UNISA College of Science, Engineering and Technology is highly appreciated. We would also like to thank the South African Institute of Physics for hosting and managing the symposium website as well as providing expert guidance in organizing the symposium. The organizers thank all the participants for making the symposium a success. The International Advisory Committee was helpful in providing valuable advice in the review of the contributions to the symposium.

The Editors
G J Rampho and M L Lekala

UNISA university of south africa

IOP Conf. Series: Journal of Physics: Conf. Series **915** (2017) 011001 doi:10.1088/1742-6596/915/1/011001

Local Organizing Committee

Lekala M L (Chair)	: University of South Africa
Rampho G J (Scientific Chair)	: University of South Africa
Adam R M	: Square Kilometer Array (South Africa)
Rakityansky S A	: University of Pretoria
Mukeru B	: University of South Africa

International Advisory Committee

J-M Richard (France)
W Plessas (Austria)
M Braun (South Africa)
H Masui (Japan)
SL Yakovlev (Russia)
L Tomio (Brazil)
EK Motovilov (Russia)
H Mahammadreza (USA)
A Karpov (Russia)
F Azaiez (South Africa)

Sponsors

- University of South Africa (UNISA)
- National Institute for Theoretical Physics (NITheP)
- South African Institute of Physics (SAIP)

List of Participants

G Rampho	: University of South Africa (South Africa)
M Lekala	: University of South Africa (South Africa)
B Mamba	: University of South Africa (South Africa)
E Meoto	: University of South Africa (South Africa)
B Mukeru	: University of South Africa (South Africa)
M Braun	: University of South Africa (South Africa)
S Rakityansky	: University of Pretoria (South Africa)
T Tshipi	: University of Pretoria (South Africa)
M Gopane	: University of Pretoria (South Africa)
P Vandrager	: University of Pretoria (South Africa)
S Oryu	: Tokyo University of Science (Japan)
L Tomio	: Universidade Estadual Paulista (Brazil)
I Lagaris	: University of Ioannina (Greece)
E Hiyama	: RIKEN (Japan)
E Kolganova	: Joint Institute for Nuclear Research (Russia)
H Masui	: Kitami Institute of Technology (Japan)
A Motovilov	: Joint Institute for Nuclear Research (Russia)
F Azaiez	: iThemba LABS (South Africa)
C Sofianos	: Sofianos Family (South Africa)

IOP Conf. Series: Journal of Physics: Conf. Series **915** (2017) 011002 doi:10.1088/1742-6596/915/1/011002

Peer review statement

All papers published in this volume of *Journal of Physics: Conference Series* have been peer reviewed through processes administered by the proceedings Editors. Reviews were conducted by expert referees to the professional and scientific standards expected of a proceedings journal published by IOP Publishing.

IOP Conf. Series: Journal of Physics: Conf. Series **915** (2017) 012001 doi:10.1088/1742-6596/915/1/012001

A new horizon of Few-Body Problems: Exact Coulomb treatment and the energy-momentum translation of the three-body Faddeev equation

S Oryu[1], T Watanabe[1], Y Hiratsuka[1], A Kodama[1], Y Togawa[2] and M Takeda[2]

[1] Dept. of Physics, Faculty of Science and Technology, Tokyo University of Science, Noda Chiba 278-8510, Japan

[2] Dept. of Information Sciences, Faculty of Science and Technology, Tokyo University of Science, Noda Chiba 278-8510, Japan

E-mail: `oryu@rs.noda.tus.ac.jp`

Abstract. We propose an off-shell Coulomb-like amplitude with an on-shell phase shift that is accurate to within 9-10 digits. The full Coulomb amplitude is separated into the leading and the auxiliary amplitudes using "the two-potential theory". The leading amplitude reproduces most of the on- and off-shell parts, while the auxiliary amplitude contributes mainly to the off-shell part and minimally to the on-shell part. We then review the three-body calculation method developed during the last four decades. A reminder to a threshold behavior investigation method is pointed out based on the three-body Faddeev equation. We discuss the Efimov physics and some extensions, which are recovered from our predictions. The method may suggest a promising technique to resolve the existing discrepancies between current experimental and theoretical values.

1. Introduction

The Coulomb phase shift renormalization method has a long history [1, 2, 3, 4, 5, 6, 7]. The most plausible method was presented by Alt et al. [8, 9, 10, 11]. Recently, we checked the reliability of the renormalization method. It was found that the phase shift can be definitely represented by the renormalization method. However, the screening range is unusually long [12]. In order to obtain the proton-proton phase shift $\sigma_0(k) = 96.8$ degrees at $E = 1$ keV by the renormalization method, we need to solve the Lippmann-Schwinger (LS) equation in momentum space with the screening range of 3 nm, 1400 Gauss points, to within 250 significant digits (see figure 1). However, a serious cancelation of significant digits occurs, resulting in converegence to a incorrect, unreliable value. Therefore, we confirmed that the phase shift renormalization is well performed by such a proper significant digits. However, Alt's method cannot treat the nuclear interaction by the usual FORTRAN program.

We presented the following generalized screened Coulomb potential $V^R(r)$ and the renormalization phase $\phi(k, R)$ [13]

$$V^R(r) = \frac{2k\eta(k)}{r} e^{-(r/R)^m} \tag{1}$$

$$\phi(k,R) = \eta(k)\Big(\ln 2kR - \gamma/m\Big). \tag{2}$$

In this case, we obtain a short cut-off potential by increasing m. After that, Deltuva et al. used an effective short-range, screened Coulomb potential, by adopting $m = 5$. In the three-body calculation, such two-body information is required for a wide two-body "sub-energy" region $-\infty < z'' \leq E$. This is because,

$$z'' = E - \frac{q''^2}{2\mu} \tag{3}$$

$$\text{for} \quad 0 \leq q'' < \infty. \tag{4}$$

In order to obtain the two-body off-shell amplitude for the screened Coulomb potential, by the LS equation, one needs zero energy. Therefore, the most remarkable region of the Coulomb off-shell amplitude is the origin i.e. $z = 0$ for the two- and three-body problems. The appropriate ranges in the energy region from 0.1 keV to 100 MeV, are listed in table 1. It is found that the $m = 5$ case yields on the average one order of magnitude shorter range than the case $m = 1$. In order to obtain higher accuracy in the phase shift, still longer ranges are required. If the renormalization method is used, efforts for getting the shorter ranges were shown to be ineffective. Despite of the suggestion of Deltuva et al., it is found that the case $m = 5$ does not essentially improve over $m = 1$ case [14, 15, 16]. Therefore, we suspect that the general use of unsatisfied Coulomb off-shell amplitude could mislead the three-nucleon scattering calculation. This may lead to desparity, considering that the ASZ method may not have been successful for the past four decades.

Table 1. Screening ranges for $m = 1$ and $m = 5$ with respect to the generalized screened Coulomb potential by (1). These ranges are obtained to guarantee the three digit accuracy in the phase shift. Present ranges belongs to the asymptotic region.

Energy [MeV]	$m = 1$	$m = 5$	σ_0 [deg.]
100	11 fm	4 fm	-0.370
50	35 fm	8 fm	-0.523
20	110 fm	17 fm	-0.826
10	300 fm	35 fm	-1.17
1	900 fm	620 fm	-3.66
0.1	1.0×10^4 fm	3.7×10^3 fm	-10.7
0.01	1.0×10^5 fm	7.7×10^4 fm	-16.3
0.001	3.0×10^6 fm	4.5×10^5 fm	96.8
0.0001	3.0×10^7 fm	6.2×10^6 fm	949.9

2. On-shell Coulomb Amplitude

In [12], we introduce an off-shell Coulomb-like amplitude. Before that, we have introduced the so-called "critical range" which satisfies zero renormalization phase [17, 18, 19, 20, 21, 22, 23, 24, 25]. In this paper, we will review our method for the on-shell Coulomb amplitude where the Coulomb phase shift is defined by the screened Coulomb phase shift plus a renormalization phase: $\sigma_0(k) = \delta_0^R(k) + \phi(k,R)$. Therefore, the on-shell scattering amplitude is separated into two parts for the S-wave case by

$$f_0^C(k) = e^{2i\phi(k,R)}\Big(\frac{e^{2i\delta_0^R(k)} - 1}{2ik}\Big) + \Big(\frac{e^{2i\phi(k,R)} - 1}{2ik}\Big). \tag{5}$$

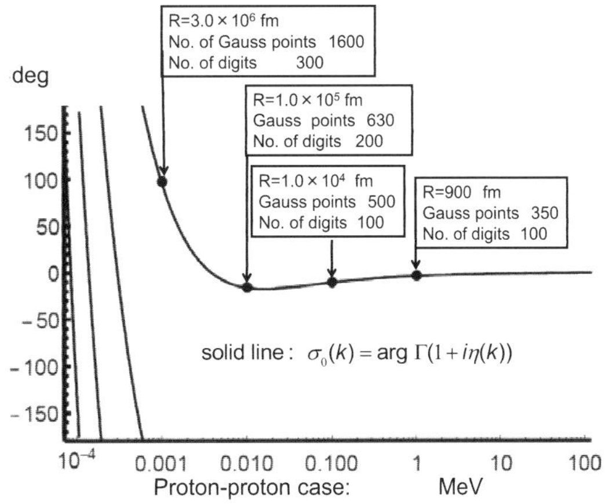

Figure 1. Necessary digits to obtain three digits accurate Coulomb phase shift by the renormalization method with the LS equation for $m = 1$ case.

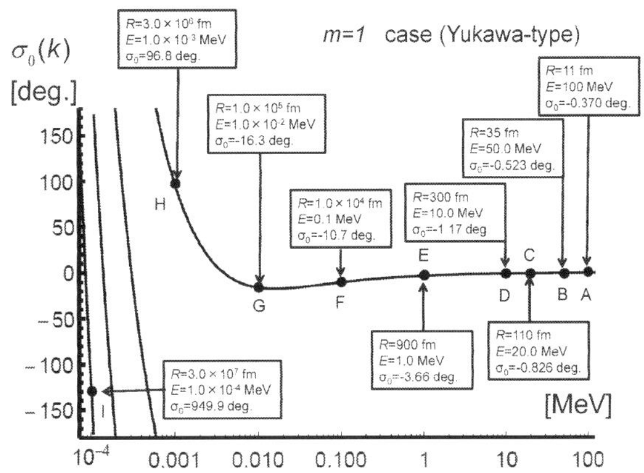

Figure 2. Necessary screening range to represent the three digits accurate Coulomb phase shift by the renormalization method. $m = 1$ case in (1).

However, the "renormalized Coulomb amplitude" is usually defined by the first term of the right hand side of (5), where the second term has been neglected without explanation [9].

In our method, we require $\phi(k, R) = n\pi$ for $n = 0, 1, 2, \cdots$. Using this criterion, a "critical range" is obtained, and the screened Coulomb phase shift becomes the Coulomb phase shift itself: $\delta_0^R(k) = \sigma_0(k)$, and the on-shell amplitude is given by $f_0^C(k) = (e^{2i\delta_0^R(k)} - 1)/2ik$. In other words, the Coulomb phase shift can be obtained directly from the critical range. We pointed out [26] that the screening range is not obtained by increasing a single range but by concatenating many discrete bands. Since Alt's renormalization phase is given by (2) in the asymptotic region [27], we may set:

$$\phi(k, R) = \eta(k)\Big(\ln 2kR - \frac{\gamma}{m}\Big) = n\pi, \qquad (6)$$

where $n\pi$ is adopted instead of $2n\pi$, because the difference corresponds to only the sign of the

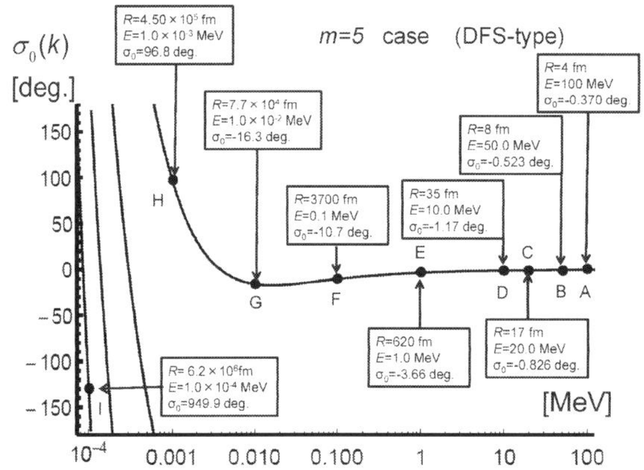

Figure 3. Necessary screening range to represent the three digits accurate Coulomb phase shift by the renormalization method. $m = 5$ case in (1).

wave function. Therefore, our "critical range", hereafter Alt's range, becomes:

$$R \;=\; \frac{1}{2k} \exp\Big(\frac{\pi}{\eta(k)} + \frac{\gamma}{m}\Big). \tag{7}$$

We conclude that the screened Coulomb phase shift becomes the Coulomb phase shift by using this critical range. However, a very long range is required to satisfy the asymptotic properties (see figures 2 and 3). Such a long range, due to the cancellation of significant digits, renders the solution of the LS equation quite hard. In order to avoid these difficulties, we obtained the best range parameters, by a fitting method at a region much shorter than the asymptotic domain where the phase shift calculation guarantees nine to ten digits accuracy (see figures 1, 2, 3 and table 1) [27]. Hereafter, the term "critical range" will be used to refer to these best range parameters.

3. A new method for the off-shell Coulomb-like amplitude

We presented an off-shell Coulomb-like amplitude [12]. In this paper, we review it and extend the results. We would like to introduce the two-potential theory in r-space. The Schrödinger equation for the Coulomb wave function may be written in operator form as:

$$(E - H_0 - V^{\mathcal{R}})|\psi^C\rangle = V^{\phi}|\psi^C\rangle = (V^C - V^{\mathcal{R}})|\psi^C\rangle. \tag{8}$$

Now, we assume a relation: $V^C|\psi^C\rangle \approx V^{\mathcal{R}'}|\psi^{\mathcal{R}'}\rangle$, where $V^{\mathcal{R}'}$ is a screened Coulomb potential of range $\mathcal{R}'$ with proper asymptotic properties and $|\psi^{\mathcal{R}'}\rangle$ the corresponding wave function. This assumption makes sense in the frame of the above discussion of Alt's Coulomb renormalization method. Therefore, the asymptotic behavior for such a long range region leads to $V^{\phi}|\psi^C\rangle \approx (V^{\mathcal{R}'} - V^{\mathcal{R}})|\psi^{\mathcal{R}'}\rangle \equiv V^{\phi'}|\psi^{\mathcal{R}'}\rangle \to 0$ (for $r \to \infty$). With this condition, we can define a general solution $|\psi^{\mathcal{R}}\rangle$ for the homogeneous equation in (8). Finally, the general solution of (8) is given by

$$|\psi^C\rangle \;=\; |\psi^{\mathcal{R}}\rangle + G^{\mathcal{R}} V^{\phi}|\psi^C\rangle \tag{9}$$

$$G^{\mathcal{R}} \;=\; [E - H_0 - V^{\mathcal{R}}]^{-1} = G_0 + G_0 T^{\mathcal{R}} G_0 \equiv G_0 \overline{\omega}^{\mathcal{R}}, \tag{10}$$

$$\overline{\omega}^{\mathcal{R}} \;=\; 1 + T^{\mathcal{R}} G_0, \tag{11}$$

$$\omega^{\mathcal{R}} \;=\; 1 + G_0 T^{\mathcal{R}}, \tag{12}$$

IOP Conf. Series: Journal of Physics: Conf. Series **915** (2017) 012001 doi:10.1088/1742-6596/915/1/012001

where $G^{\mathcal{R}}$ is Green's function with the screened Coulomb potential $V^{\mathcal{R}}$.

By multiplying (9) from the left with V^{ϕ}, an amplitude $t^{\phi\mathcal{R}}$ may be defined using $|\psi^{\mathcal{R}}\rangle$ and the plane wave $|\psi_0\rangle$, as:

$$V^{\phi}|\psi^{C}\rangle \equiv t^{\phi\mathcal{R}}|\psi^{\mathcal{R}}\rangle = t^{\phi\mathcal{R}}\omega^{\mathcal{R}}|\psi_0\rangle\,. \tag{13}$$

Therefore, $t^{\phi\mathcal{R}}$ introduces all the long range properties in the Coulomb wave function, and also merits for the proper asymptotic behavior mentioned above. Therefore, (9) is rewritten by using (10), and (13),

$$|\psi^{C}\rangle = |\psi^{\mathcal{R}}\rangle + G^{\mathcal{R}}t^{\phi\mathcal{R}}|\psi^{\mathcal{R}}\rangle = |\psi^{\mathcal{R}}\rangle + G_0\overline{\omega}^{\mathcal{R}}t^{\phi\mathcal{R}}\omega^{\mathcal{R}}|\psi_0\rangle \tag{14}$$

$$= |\psi^{\mathcal{R}}\rangle + G^{\mathcal{R}}V^{\phi}|\psi^{C}\rangle = |\psi^{\mathcal{R}}\rangle + G_0\overline{\omega}^{\mathcal{R}}V^{\phi}|\psi^{C}\rangle. \tag{15}$$

Multiplying (14) from the left with V^{ϕ}, and using (13), yields:

$$V^{\phi}|\psi^{C}\rangle = t^{\phi\mathcal{R}}|\psi^{\mathcal{R}}\rangle = V^{\phi}|\psi^{\mathcal{R}}\rangle + V^{\phi}G^{\mathcal{R}}t^{\phi\mathcal{R}}|\psi^{\mathcal{R}}\rangle. \tag{16}$$

Here, we define a Coulomb-like off-shell T-matrix $\tilde{T}^{C}$ by

$$G_0^{-1}|\psi^{C}\rangle = V^{C}|\psi^{C}\rangle \equiv \tilde{T}^{C}|\psi_0\rangle, \tag{17}$$

$$G_0^{-1}|\psi^{\mathcal{R}}\rangle = V^{\mathcal{R}}|\psi^{\mathcal{R}}\rangle = T^{\mathcal{R}}|\psi_0\rangle, \tag{18}$$

where $\tilde{T}^{C}$ has all the properties of the long range Coulomb wave function and the potential. In addition it satisfies the Coulomb asymptotic behavior and our requirement of (6) which is equivalent to the *Lemma* stated in our recent article [26].

Multiplying (14) from the left with V^{C} and using (17), (18), we obtain

$$\tilde{T}^{C}|\psi_0\rangle = T^{\mathcal{R}}|\psi_0\rangle + \overline{\omega}^{\mathcal{R}}t^{\phi\mathcal{R}}\omega^{\mathcal{R}}|\psi_0\rangle \tag{19}$$

$$= T^{\mathcal{R}}|\psi_0\rangle + \overline{\omega}^{\mathcal{R}}V^{\phi}|\psi^{C}\rangle. \tag{20}$$

Therefore, (19), (16), and (10) lead to the following operator relations:

$$\tilde{T}^{C} = T^{\mathcal{R}} + T^{\phi\mathcal{R}} = T^{\mathcal{R}} + \overline{\omega}^{\mathcal{R}}t^{\phi\mathcal{R}}\omega^{\mathcal{R}}, \tag{21}$$

$$t^{\phi\mathcal{R}} = V^{\phi} + V^{\phi}G^{\mathcal{R}}t^{\phi\mathcal{R}} \tag{22}$$

$$T^{\mathcal{R}} = V^{\mathcal{R}} + V^{\mathcal{R}}G_0T^{\mathcal{R}}, \tag{23}$$

where (21) is known as the amplitude by the two-potential theory. Multiplying from the left (20) with $\langle\psi_0|$ and using $\langle\psi_0|\overline{\omega}^{\mathcal{R}} - \langle\psi^{\mathcal{R}}|$, we obtain

$$\langle\psi_0|\tilde{T}^{C}|\psi_0\rangle = \langle\psi_0|T^{\mathcal{R}}|\psi_0\rangle + \langle\psi_0|\overline{\omega}^{\mathcal{R}}V^{\phi}|\psi^{C}\rangle$$

$$= \langle\psi_0|T^{\mathcal{R}}|\psi_0\rangle + \langle\psi^{\mathcal{R}}|V^{\phi}|\psi^{C}\rangle. \tag{24}$$

(24) is equivalent to (21), and the amplitude $T^{\phi\mathcal{R}}$ is obtained by using the partial wave functions $|\psi_L^{\mathcal{R}}\rangle$ and $|\psi_L^{C}\rangle$.

$$T_L^{\phi\mathcal{R}}(p,p';E) \equiv \langle\psi_L^{\mathcal{R}}|V^{\phi}|\psi_L^{C}\rangle = \int_0^{\infty}\psi_L^{\mathcal{R}*}(p,r)V^{\phi}(r,k)\psi_L^{C}(p',r)r^2dr. \tag{25}$$

This result guarantees the existence of the off-shell Coulomb-like T-matrix by (21) and (25), which was not clearly stated in [28]. The off-shell value given by (25), is calculated numerically, where the integrand is highly oscillatory and vanishes for very large values of r. Therefore, (25) contributes to off-shell, and half-on-shell values.

In our method, the on-shell value of $T^{\mathcal{R}}$ in (21) gives 9–10 digits of accuracy [27]. Therefore, it is obvious that the on-shell value of $T^{\phi\mathcal{R}}$ contains only the next order correction need to obtain an accuracy of 9–10 digits in the phase shift. However, the numerical calculation of the on-shell value of (25) is very sensitive to the convergence condition. Therefore, in order to calculate the on-shell value of (25), we can adopt a screened Coulomb potential with a very large range ($\mathcal{R} \ll$)$\mathcal{R}'$ which appears in the asymptotic region and is introduced to obtain the solution of (8). For such an $\mathcal{R}'$, we can accomplish the following relation:

$$T_L^{\phi\mathcal{R}}(k,k;E) = \int_0^\infty \psi_L^{\mathcal{R}*}(k,r)V^{\phi}(r,k)\psi_L^{C}(k,r)r^2dr \equiv \Delta_L \tag{26}$$

$$\rightarrow \int_0^\infty \psi_L^{\mathcal{R}*}(k,r)V^{\phi'}(r,k)\psi_L^{\mathcal{R}'}(k,r)r^2dr \cong -\frac{|T_L^{\mathcal{R}}|^2}{\mathcal{I}m(T_L^{\mathcal{R}})} \times \phi'_L(k) \tag{27}$$

with

$$\phi'_L(k) \equiv \sigma_L(k) - \delta_L^{\mathcal{R}}(k) = \tan^{-1}\frac{\mathcal{I}m(T_L^{\mathcal{R}}+\Delta_L)}{\mathcal{R}e(T_L^{\mathcal{R}}+\Delta_L)} - \tan^{-1}\frac{\mathcal{I}m(T_L^{\mathcal{R}})}{\mathcal{R}e(T_L^{\mathcal{R}})}, \tag{28}$$

where the last term in (27) is an auxiliary amplitude, $\phi'_L(k)$ is the discrepancy between the analytic Coulomb phase shift and our direct phase shift. We can find an adequate $\mathcal{R}'$ which satisfies (27). In [12], we mentioned that the value of (26) vanishes. However, such a condition is satisfied only for the case where the eigen-values of $V^C|\psi^C\rangle$ and $V^{\mathcal{R}}|\psi^{\mathcal{R}}\rangle$ are identical. Our fitting procedure albeit not exact, reproduces the Coulomb phase shift with an accuracy of 9–10 digits. Therefore, we can conclude that the first term $T^{\mathcal{R}}$ represents almost all on- and off-shell Coulomb amplitude, with a compensation coming from the off-shell part of the second term.

4. Review of the three-hadron problem

Half a century has passed since the Faddeev equation was presented [29]. In the beginning, it seemed that the obtained results fitted the experimental data well, in contrast to the results of other methods. After that, improved results have been mainly due to the rapid growth of the computer capacity as to how many partial waves can be incorporated, and what kind of realistic two-body and three-body forces can be treated. These efforts have been done somewhat with a technical professionalism. However, the ideas of the theoretical and methodological approach of the Faddeev equation were investigated only in the beginning, that is, in 1960s and 1970s [30, 31, 32, 33, 34]. It seems that today, is a good time to start a "deep critical rethinking" of the results obtained so far by the usual Faddeev calculations, in order to understand whether the fundamental problems of the Faddeev equation still exist or not.

First of all, the quantitative agreements for A_y are not yet satisfied [35]. Recently, the intensive discussion about the disagreements for the cross section minimum and A_y maximum were pointed out by Sagara [36]. The so-called "Sagara discrepancy" was illustrated by a ratio between theoretical and experimental data in the cross section minimum and the maximum of A_y: $S_{\rm disc}$ ≡(calculation-experiment)/experiment [37, 38, 39]. Such discrepancies have an energy dependence which is commonly inversely proportional to the energy. It was claimed that the cross section minimum problem was solved by using a three-body force named 2π 3NF [40]. However, it seems that the A_y puzzle has not yet been solved by such a systematic way with 2π 3NF and some other intensive 3NFs [41]. We are concerned about the energy dependence of the Sagara discrepancy not only for the A_y puzzle, but also the cross section minimum. If such a 3NF can fundamentally reproduce the cross section minimum, the A_y-puzzle should also be solved by the 3NF at the same time. However, the puzzle is not solved yet. The concern arises from whether the 3NF is fundamental or an ad hoc remedy in the three-nucleon problem.

Another problem discussed but remained unsolved is the break-up problem in the case of "the Space star anomaly", although 3NF and many different so-called realistic potentials were used. Also an incomplete differential cross section could not yet be represented by the Faddeev calculation [30]. On the other hand, the discrepancy in the triton binding energy between the experimental data is compensated by a three-body force [42]. It was found that the doublet n-d scattering length ($a^2_{\rm nd}$) is closely related to the three-body force [43]. The closed relation between $a^2_{\rm nd}$ and 3NF is reminiscent of the relation between the "n-d threshold" and the 3NF. Many attempts are carried out by using realistic potentials and corresponding three-body forces, relativistic effects, the Coulomb interaction, new degrees of freedom etc. However, those methods could not satisfactorily explain the above discrepancies. The three-body force is mainly an isotropic interaction. Furthermore, the renormalization phase of the Coulomb amplitude is also isotropic. Such isotropic interactions can not affect the P-wave dominant A_y value, and the effect of the isotropic interaction could work somewhat negatively. Therefore, we suspect that the energy dependence in the Sagara discrepancy could be closely related to an anisotropic energy-momentum usage or kinematics in the Faddeev equation.

The second is the three-body bound state which is obtained by solving the two-variable eigen-equation as the three-body Faddeev equation. However, the two-variable eigen-equation is "not well defined" mathematically, although the one variable eigen-equation is well defined [44]. Therefore, we should separate the two-variable three-body Faddeev equation into one-variable two-body LS equations, and a one-variable quasi-two-body multi-channel LS equation. Each of them gives eigen-values $\epsilon^{(2)}$ for $(bc) \to b+c$ and ζ by solving one-variable eigen-equations, where we obtain a separation energy ζ for the reaction $(abc) \to (bc) + a$ by the quasi-two-body multi-channel LS equation. Therefore, the three-body binding energy is obtained by $\epsilon^{(3)} = \zeta + \epsilon^{(2)}$ [45]. This method coincides with the Jacobi-coordinate anti-symmetrized molecular dynamics, where bound colonies are grown up sequentially from many free-particle groups, by cooling the system in a timely manner, and not instantaneously as in the usual Faddeev method [46, 47, 48, 49, 50].

4.1. Three-body kinematics

Now, let us start with the well-known three-body kinematics for the scattering with a incident particle and a target. The incident particle has a kinetic energy E_{Lab} with the momentum q_L,

$$E_{Lab} = \frac{q_L^2}{2m} = \frac{q_L^2}{2(m+M_t)} + E_{\rm cm} = \Big(\frac{m}{m+M_t}\Big) E_{Lab} + E_{\rm cm} \tag{29}$$

$$E = E_{\rm cm} - \epsilon_B = \frac{q^2}{2\mu} + \frac{p^2}{2\nu}, \quad \text{(target binding energy} = \epsilon_B) \tag{30}$$

$$E_{\rm cm} = \frac{\overline{q}^2}{2\mu} + \frac{\overline{p}^2}{2\nu}, \quad \text{(target unbound system)}, \tag{31}$$

where m and M_t are the masses of the incident particle and the target. μ denotes the reduced mass between a pair and the spectator particle, and ν is the reduced mass in the pair, respectively. Therefore, q is the relative momentum between the pair and the spectator, and p is the relative momentum in the pair after three-body break-up. $E_{\rm cm}$ is the three-body incident center of mass energy, and ϵ_B is the target binding energy. Therefore E is the three-body free energy after the deuteron break-up. If there are no two-body sub-bound states in the intermediate state, then the three-body total energy $E_{\rm cm}$ should be used, where one of the three particles, at least, is virtual.

Usually, the Faddeev equation and the AGS equation commonly use the three-body free energy E in the Green's function: $G_0(E)$ where real-three-particles are transferred in between two different three-body channels above the break up threshold. However, if the initial and final channels are deuteron bound states in the Born term, the spectator particles gain the

IOP Conf. Series: Journal of Physics: Conf. Series **915** (2017) 012001 doi:10.1088/1742-6596/915/1/012001

deuteron binding energy. Therefore, the initial and the final state's kinetic energies are given by $E_{\rm cm} = \overline{q}^2/2\mu = \overline{q}'^2/2\mu$. In the Born term of AGS equation, the Green's function could be written by $G_0(\overline{q},\overline{q}';E)$. Therefore, a transferred momentum in the Green's function could be partly virtual, because all the particle-lines in the Feynman diagram are "outer-lines" where the energy-momentum should be conserved. While, if we choose $E_{\rm cm}$ instead of E in the Green's function, the three-particle momenta are all real by (31) and (36) with $G_0(\overline{q},\overline{q}';E_{\rm cm})$. $E_{\rm cm}=0$ indicates the quasi-two-body threshold (Q2T) in the three-body scattering which causes a divergence of the two-body propagator $\tau(z)$. The n-d scattering length is calculated with such analytic properties as to $E_{\rm cm}=0$.

On the other hand, in the intermediate three-body lines as the "inner lines" of the Feynman diagram, we are concerned about two-type of three-body energies: the three-body energy $E_{\rm cm}$ before the break up, and the three-body energy E after the break up. It is found that by an energy-momentum translation:

$$E \to E_{\rm cm} = E + \epsilon_B \tag{32}$$

$$q^2/2\mu \to \overline{q}^2/2\mu = q^2/2\mu + \epsilon_B, \tag{33}$$

$$\overline{p}^2/2\nu = p^2/2\nu, \tag{34}$$

we obtain, for the usual Born term in the three-identical particle system,

$$Z_{\alpha,\beta}(q,q';E) = \frac{g_\alpha(p)g_\beta(p')(1-\delta_{\alpha\beta})}{\Big(E+\epsilon_B\Big) - \Big(q^2/2\nu_\beta + q'^2/2\nu_\alpha + qq'x/m + \epsilon_B\Big)} \tag{35}$$

$$= \frac{g_\alpha(\overline{p})g_\beta(\overline{p}')(1-\delta_{\alpha\beta})}{E_{\rm cm} - \overline{q}^2/2\nu_\beta - \overline{q}'^2/2\nu_\alpha - \overline{q}\overline{q}'\overline{x}/m} = Z_{\alpha,\beta}(\overline{q},\overline{q}';E_{\rm cm}). \tag{36}$$

where $x = \mathbf{q}\mathbf{q}'/qq'$ and $\overline{x} = \overline{\mathbf{q}\mathbf{q}}'/\overline{qq}'$. The propagator becomes,

$$\tau_\gamma(E - q'^2/2\mu) = \tau_\gamma([E+\epsilon_B] - [q'^2/2\mu + \epsilon_B]) = \tau_\gamma(E_{\rm cm} - \overline{q}'^2/2\mu). \tag{37}$$

Therefore, the energy-momentum translation does not affect the function itself, however the integral may be modified due to the prolonged range: $0 \le \overline{q}^2/2\mu \le \epsilon_B$ or $-\epsilon_B \le q^2/2\mu \le 0$.

4.2. Non-Efimov but Efimov-like states

One author (S.O.) pointed out that $E_{\rm cm}=0$ in (36) for NNπ system, generates a singularity-coincidence between the Born term and the propagator $\tau_\gamma(E_{\rm cm} - \overline{q}'^2/2\mu)$ [51]. This coincidence is very similar to the Coulomb scattering in the LS equation where the singularity of the Coulomb potential and of the propagator occurs simultaneously. Since our potential singularity in the present case arises at $E_{\rm cm}=0$, then we carried out an energy average around $E_{\rm cm}=0$ after the Fourier transformation of (36). Finally, we obtained $V(r) = V_0 a^{2\gamma+2}/[r(r/2+a)^{2\gamma+2}]$-type potentials with parameters a and γ and a depth constant $V_0(<0)$ for the lowest order term in the local potential expansion of the Born term. Therefore, we obtained $V_0(2a)/r^2$ potential for $\gamma=-1/2$, and Van der Waals potentials $V_0(2a)^5/r^6$ for $\gamma=3/2$, $V_0(2a)^6/r^7$ for $\gamma=2$ potentials in the long range region for $a \ll r$. While in the shorter range region $r \ll a$, they become Yukawa-type potentials $V_0 e^{-r/2a}/r$, $V_0 e^{-5r/2a}/r$, and $V_0 e^{-6r/2a}/r$, respectively (see table 2). Beside the lowest order, the higher order terms affect the shorter range region adding to the Yukawa-type potential. Therefore, the shorter range region could be strongly modified from the Yukawa-type potential near the core region, however our longer range predicted potential can not be changed by the higher order terms.

These potentials are closely related to the mass ratio between a transfer particle and the parent particles, which can be easily understood by the reason that if the massless particle is

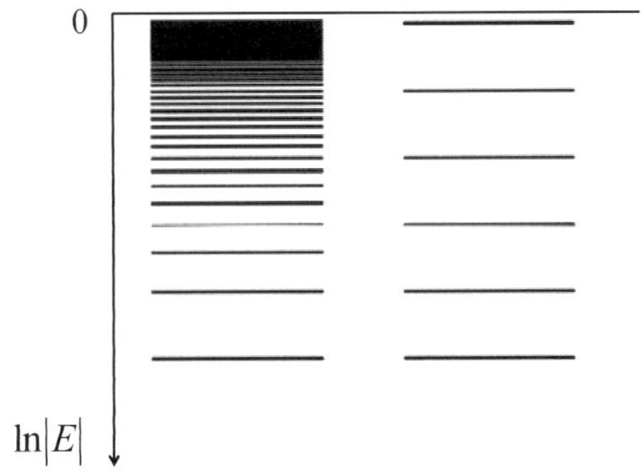

Figure 4. Bound state energy levels for the potentials $1/r$ (left) and $1/r^2$ (right) where the ground state energies are fitted. The vertical unit is arbitrary but the logarithmic unit of the absolute value is taken.

transferred, then the potential will be $1/r$ which is given by $\gamma = -1$. It is interesting that $\gamma = -1$ or a massless particle creates a unique potential V_0/r where two-type of potentials for $a \ll r$ and $r \ll a$ cross each other (see table 2). The above discussion, that two types of potentials, a long range and a short range, arise from a particle-transfer mechanism, seems to be a universal structure in physics. If there is a counterargument about our "energy average" method, we can directly solve the three-body equation with (36), where we need a special technique to treat it in the neighborhood of the singular region. Finally, we can also obtain bound states near the Q2T or the energy dependent two-body quasi (E2Q)-threshold.

Table 2. An effective potential $V_0\, a^{2\gamma+2}/[r(r/2+a)^{2\gamma+2}]$ is illustrated, which is given by an energy average around the Q2T (or $E_{\rm cm} = 0$) with two-parameters a and γ. The potential properties for the longer and shorter ranges are shown with respect to the parameter γ. $V_0(<0)$ is the potential depth given analytically by the Born term of the form (36).

γ	short range potential $r \ll a$	potential	long range potential $a \ll r$
-1	V_0/r	V_0/r	V_0/r
-1/2	$V_0 e^{-r/2a}/r$	$V_0(2a)/[r(r+2a)]$	$V_0(2a)/r^2$
0	$V_0 e^{-2r/2a}/r$	$V_0(2a)^2/[r(r+2a)^2]$	$V_0(2a)^2/r^3$
1/2	$V_0 e^{-3r/2a}/r$	$V_0(2a)^3/[r(r+2a)^3]$	$V_0(2a)^3/r^4$
1	$V_0 e^{-4r/2a}/r$	$V_0(2a)^4/[r(r+2a)^4]$	$V_0(2a)^4/r^5$
3/2	$V_0 e^{-5r/2a}/r$	$V_0(2a)^5/[r(r+2a)^5]$	$V_0(2a)^5/r^6$
2	$V_0 e^{-6r/2a}/r$	$V_0(2a)^6/[r(r+2a)^6]$	$V_0(2a)^6/r^7$
.	...	...	...
.	...	...	...

It is well known that $1/r^2$ potential generates the Efimov-like states, although our case does not require the infinite number of the two-body scattering length: $a_{\rm scatt} \neq \pm\infty$. However, when the two-body binding energy becomes zero then $0 = E_{\rm cm} = E + \epsilon_B \to E$ which is the case of Efimov states [52, 53]. This is also the case that the energy-momentum translation is denied. In this sense, our prediction covers the Efimov states. We can say that our prediction at the Q2T is generalized to other thresholds such as three-body, four-body, five-body, ..., thresholds [54]. This fact suggests that "observed and unobserved energy levels" in any many-body system give rise to a long range potential near the threshold. We would like to suggest that these levels could be a key which can "control the nuclear reactions" where the long range attractive nuclear interaction could interfere with the Coulomb repulsive force in the pico-size region [51].

5. Conclusion and Discussion

The reliability of the Coulomb renormalization method is confirmed. However, its application to the three-body problem is almost impossible because the screening range is too long. When Alt-type Coulomb renormalization is used, a serious cancelation of significant digits occurs in the two-body LS and in the three-body Faddeev equations. This cancelation causes the method to converge, without any warning, to an erroneous value, hence rendering the calculation unreliable. We present a shorter ranged screened Coulomb potential which can reproduce the Coulomb phase shift to within 9–10 digits accuracy. Furthermore, the off-shell Coulomb-like amplitude is proposed by the two-potential method in r-space representation where the off-shell amplitude can be calculated numerically. By our theory, the Coulomb treatment in the three-body problem could be enormously improved for the first time in a period of half a century. Finally, we would like to emphasize that the method is very general, and it can treat all kinds of charged particle systems, from the electron-electron to the heavy ion-heavy ion, using the universal variables.

We reviewed the discrepancy between the experimental data and the usual Faddeev calculated results. Sagara discrepancy is the most important admonition for this problem. We also added some other problems that seemed solvable by incorporating the three-body force. We pointed out that the Q2T is important for improving the Faddeev treatment. One of the examples was the case of the NNπ system, and we presented a possible existence of a local long-range NN$'$ potential around Q2T, and NN$'$ scattering length and πD one [54]. This potential results to some interesting bound states, and to the existence of long-range potentials such as the $1/r^2$ type and the Van der Waals potential. The $1/r^2$ potential gives infinite number of bound states below the Q2T, without Efimov's requirement for infinite scattering length of two-body sub-system. Therefore, our theory predicts, in addition to the Efimov states, some other interesting aspects of nuclear physics.

Acknowledgments
The authors would like to pay tribute to the memory of the late Professor S. A. Sofianos, for his support and continuous encouragement since the very beginning of this idea.

References

[1] Tobocman W and Kalos M H 1955 Numerical Calculation of (d,p) Angular Distributions *Phys. Rev.* **97** 132
[2] Veselova A M 1970 Separation of two-particle Coulomb singularities in a system of three charged particles (in Russian) *Teor. Mat. Fiz.* **3** 326
[3] Veselova A M 1970 Separation of two-particle Coulomb singularities in a system of three charged particles *Theor. Math. Phys.* **3** 542
[4] Veselova A M 1972 Determination of scattering amplitudes in problems with two and three charged particles (in Russian) *Teor. Mat. Fiz.* **13** 368
[5] Veselova A M 1972 Determination of scattering amplitudes in problems with two and three charged particles *Theor. Math. Phys.* **3** 1200
[6] Veselova A M 1978 Integral equations for three particles with Coulomb long-range interaction (in Russian) *Teor. Mat. Fiz.* **35** 180
[7] Veselova A M 1978 Integral equations for three particles with Coulomb long-range interaction (in Russian). *Theor. Math. Phys.* **35** 395

IOP Conf. Series: Journal of Physics: Conf. Series **915** (2017) 012001 doi:10.1088/1742-6596/915/1/012001

[8] Alt E O, Sandhas W, Zankel H and Ziegelmann H 1976 Coulomb Corrections in Proton-Deuteron Scattering *Phys. Rev. Lett.* **37** 1537
[9] Alt E O, Sandhas W and Ziegelmann H 1978 Coulomb effects in three-body reactions with two charged particles *Phys. Rev.* C **17** 1981
[10] Alt E O and Sandhas W 1980 Scattering amplitudes and integral equations for the collision of two charged composite particles *Phys. Rev.* C **21** 1733
[11] Alt E O, Sandhas W and Ziegelmann H 1985 Calculation of proton-deuteron phase parameters including the coulomb force *Nucl. Phys.* A **445** 429
[12] Oryu S, Watanabe T, Hiratsuka Y and Togawa Y 2017 A Coulomb-like Off-Shell T-matrix with the Correct Coulomb Phase Shift *Few-Body Syst.* **58** 95
[13] Oryu S, Nishinohara S, Sonaoda K, Shiiki N, Togawa Y and Chiba S 2005 A novel three-charged-particle Faddeev type equation in momentum space, *AIP Conf. Proc.* **768** 433-435
[14] Deltuva A, Fonseca A C and Sauer P U 2005 Momentum-space description of three-nucleon breakup reactions including the Coulomb interaction *Phys. Rev.* C **72** 054004
[15] Deltuva A, Fonseca A C, Kievsky A, Rosati S, Sauer P U and Viviani M 2005 Benchmark calculation for proton-deuteron elastic scattering observables including the Coulomb interaction *Phys. Rev.* C **71** 064003
[16] Deltuva, A, Fonseca A C and Sauer P U 2005 Momentum-space treatment of the Coulomb interaction in three-nucleon reactions with two protons *Phys. Rev.* C **71** 054005
[17] Hiratsuka Y, Oryu S and Gojuki S 2013 D(p,p)D elastic scattering with rigorous Coulomb treatment *J. Phys G: Nucl. Part. Phys.* **40** 025106
[18] Oryu S, Hiratsuka Y, Nishinohara S and Chiba S 2012 Proton-proton phase shifts calculations in momentum space by a rigorous Coulomb treatment *J. Phys. G: Nucl. Part. Phys.* **39** 045101
[19] Oryu S, Nishinohara S, Shiiki N and Chiba S 2007 Coulomb phase shift calculation in momentum space *Phys. Rev.* C **75**, 021001
[20] Nishinohara S, Chiba S and Oryu S 2007 The Coulomb scattering in momentum space for few-body systems *Nucl. Phys.* A **790** 277c
[21] Oryu S 2006 Two- and three-charged particle nuclear scattering in momentum space: A two-potential theory and a boundary condition model *Phys. Rev.* C **73** 054001
[22] Oryu S 2007 Erratum: Two- and three-charged particle nuclear scattering in momentum space: A two-potential theory and a boundary condition model *Phys. Rev.* **76** 069901
[23] Nishinohara S 2006 *A study of Coulomb scattering by the momentum representation*, Doctoral thesis (in Japanese), (Japan: Tokyo University of Science)
[24] Hiratsuka Y 2014 *A study of the rigorous solution for two- and three-charged nucleons scattering in momentum space* (in Japanese), Doctoral thesis (Japan: Tokyo University of Science)
[25] Oryu S, Hiratsuka Y and Watanabe T 2016 A New Treatment Below the Three-Body Break up Threshold in the NNπ System *EPJ Web of Conf.* **113** 08014
[26] Watanabe T, Hiratsuka Y and Oryu S 2016 Effective Screened Coulomb Potential from e^-–e^- to ^{208}Pb–^{208}Pb Systems *EPJ Web of Conf.* **113** 03018
[27] Watanabe T, Y. Hiratsuka, Oryu S and Togawa Y 2017, A New Feature of the Screened Coulomb Potential in Momentum Space *Few-Body Syst.* **58** 80
[28] H. van Haeringen 1985 *Charged particle interactions: Theory and Formulas* (Netherlands: Coulomb Press Leyden)
[29] Faddeev L D 1960 *Zh. Eksp. Theor. Fiz.* **39** 1459; 1961 *Sov. Phys. -JETP* **12** 1014
[30] Aaron R and Amado R D 1966 Theory of the Reaction n+d $\rightarrow$ n+n+p. *Phys. Rev.* **150** 857
[31] Lovelace C 1964 Practical Theory of Three-Particle States. I. Nonrelativistic. *Phys. Rev.* **135** B1225
[32] Alt E O, Grassberger and Sandhas W 1967 Reduction of the three-particle collision problem to multi-channel two-particle. Lippmann-Schwinger equations *Nucl. Phys.* B **2** 167
[33] Cahill R T and Sloan I H 1971 Theory of neutron-deuteron break-up at 14.4 MeV. *Nucl. Phys.* A **165** 161
[34] Ebenhöh W 1972 The n+d break-up reaction with separable potentials. *Nucl. Phys.* **191** 97
[35] Koike Y and Heidenbauer J 1987 The neutron-deuteron elastic scattering with the Paris potential. *Nucl. Phys.* A **463** 365c
[36] Sagara K 2010 Experimental Investigations of Discrepancies in Three-Nucleon Reactions *Few-Body Syst.* **48** 59-108
[37] Takemiya N 1985 On the low energy n-d scattering *Prog. Theor. Phys.* **74** 301
[38] Takemiya N 1991 Analyzing powers in elastic nucleon-deuteron scattering. *Prog. Theor. Phys.* **86** 975
[39] Koike Y and Ishikawa S 1998 Possible origin of Sagara discrepancy in the cross section minimum of $p-d$ elastic scattering. *Nucl. Phys.* A **631** 683c
[40] Witała H, et al. 1998 Cross section minima in elastic $N-d$ scattering: possible evidence for three-nucleon force effects. *Phys. Rev. Lett.* **81** 1183

[41] Witała H, Golak J, Skibinski R, Topolnichki K, Kamada H, Epelbaum E, Glöckle W, Krebs H, Polyzou W N and Nogga A 2013 Calculations of Three-Nucleon Reactions. *Few-Body Syst.* **54** 897
[42] Ishikawa S, Sasakawa T, Sawada T and Ueda T 1984, Contribution of Three-Nucleon Potential to Triton Binding Energy. *Phys. Rev. Lett.* **53** 1877
[43] Oryu S and Yamada H 1994 Effects of the three-body force in three-nucleon system *Phys. Rev* C **49** 2337
[44] Furutani Kenro, Private communication.
[45] Oryu S, Hiratsuka Y and Watanabe T 2016 Few-Body Problem in Nuclear Reactions: Beyond the horizon of the three-body Faddeev equations *EPJ Web of Conf.* **122**
[46] Watanabe T and Oryu S 2006 A New Antisymmetrized Molecular Dynamics Approach to Few-Nucleon Systems *Prog. Theor. Phys.* **116** 429-434
[47] Watanabe T, Oosawa M, Saito K and Oryu S 2009 A new molecular dynamics calculation and its application to the spectra of light and strange baryons. *J. Phys. G: Nucl. Part. Phys.* **36** 015001
[48] Watanabe T, Oryu S and Saito K 2009 New approach for few-body systems based on jacobi-coordinate basis antisymmetrized molecular dynamics from atoms to quark-systems *Mod. Phys. Lett.* A **24** 797-803
[49] Taniguchi Y, Watanabe T, Sawado N and Oryu S 2004 A systematic study of light nuclei 2¡A¡8 using the AMD-method with realistic Paris AV14 nucleon-nucleon potentials *Proc. of the 17th IUPAP Conf. on Few-Body Problems in Physics* S198-S200.
[50] Watanabe T 2006 *Calculation of a few nucleons systems by antisymmetrized molecular dynamics*, Doctoral thesis (in Japanese) (Japan: Tokyo University of Science)
[51] Oryu S 2012 Universal structure of the three-body system. *Phys. Rev.* C **86** 044001
[52] Efimov V 1970 Energy levels arising from resonant two-body forces in a three-body system. *Phys. Lett.* B **33** 563
[53] Efimov V 1973 Energy levels of three resonantly interacting particles. *Nucl. Phys.* A **210** 157
[54] Oryu S, Hiratsuka Y and Watanabe T 2016 Few-Body Problem in Nuclear Reactions: Beyond the horizon of the three-body Faddeev equations *EPJ Web of Conf.* **122** 08001

Scaling behavior of scattering observables for three-body systems near the unitary limit

Lauro Tomio[1,2], M A Shalchi[1], M T Yamashita[1], M R Hadizadeh[3,4] and T Frederico[2]

[1]Instituto de Física Teórica, Univ. Estadual Paulista (UNESP), 01140-070, São Paulo, SP, Brazil (E-mail: tomio@ift.unesp.br)
[2]Instituto Tecnológico de Aeronáutica, DCTA,12228-900, São José dos Campos, Brazil
[3]Institute for Nuclear and Particle Physics and Department of Physics and Astronomy, Ohio University, Athens, OH 45701, USA
[4]College of Science and Engineering, Central State University, Wilberforce, OH 45384, USA

Abstract. This contribution reports recent investigations on low-energy scaling properties of three-body systems, by considering elastic $s-$wave collisions of a particle in a bound-state formed by the remaining two-body system. First, some previous results for the case of the halo nucleus ^{20}C will be revised, for the neutron$-^{19}$C scattering properties near the critical condition for the occurrence of an excited bound state in ^{20}C, within a neutron$-$neutron$-^{18}$C configuration. In our approach, we consider the Faddeev formalism with renormalized zero-range two-body interactions. Next, by considering the actual possibilities for verification of low-energy scaling properties in cold-atom laboratories, the approach is extended to atomic strongly-mass-imbalanced three-body systems, with two identical heavy particles and a light one. In this case, we consider that the heavy particle is being scattered by the light-heavy weakly-bound dimer. Our preliminary results for scattering observable are evidencing the universal scaling features.

1. Introduction

First shown by Efimov [1] for the case that we have three-identical particles in the unitary limit (more specifically, when the absolute value of the two-body scattering length $|a|$ becomes infinity), a discrete geometrical scaling emerges in the corresponding bound-state energy spectrum, which has an infinite number of three-body levels. Considered as a quantum mechanics pathology in the beginning [2], the Efimov effect was demonstrated in [3] to arise from essentially the same singularity structure of the kernel of the scattering equation, which is also responsible for the Thomas collapse [4] of the three-body ground state when the two-body range r_0 is reduced to zero. As shown in Ref. [3], the trace of the kernel will diverge as $\ln(|a|/r_0)$, allowing the Efimov and Thomas effects to be described in a unified way. The Efimov effect has been discussed occasionally over the years in the nuclear and atomic-molecular physics context [5–13] as related to some universal aspects of three-body quantum systems. The experimental observation of this effect in different ultracold laboratories [14–18], which was possible in view of the quantum-mechanical control of the two-body interactions provided by Feshbach resonance mechanisms [19], attracted a lot of new theoretical and experimental investigations on low-energy quantum few-body systems. Also, the existence of the long-standing prediction of an excited Efimov state in the helium trimer [9] was reported recently in Ref. [20]. As a summary of the research activity on this topic, in atomic and nuclear physics, we should mention as relevant some previous reviews, as Refs. [21–24]. In particular, for more recent ones, we can mention Ref. [25] when considering universal aspects in light halo nuclei; and Refs. [26,27] for the case of ultracold quantum gases. For a simplified pedagogical description of the Efimov effect, we can suggest Ref. [28], where this effect is shown to be a manifestation of a well-known quantum mechanics anomaly [29].

The investigations of Efimov related effects in the scattering region is quite appealing in view of recent interest in three-body systems $\alpha - \alpha - \beta$ with non-identical masses, $m_\alpha \neq m_\beta$. In nuclear physics, the actual interest goes to weakly-bound halo-nuclei systems, with two neutrons (n) in the halo and a core (c), where one can consider the scattering of a neutron by the $n - c$ weakly-bound sub-system. With increasing possibilities to be studied experimentally, we have ultra-cold atomic systems where one heavy atom is colliding with low-energy to a weakly-bound dimer formed by the remaining two particles.

Before examining the scattering region, it is first relevant to remind that in the unitary limit, it was demonstrated that two levels of the three-body Efimov spectrum are related by an exponential scaling factor given by $\exp(2\pi/s_0)$, where s_0 is a constant that varies according to the mass-ratio m_α/m_β [23]. In the case of an identical-mass system, we have the maximum energy-ratio predicted to be ~ 515, such that it will be quite difficult to be experimentally verified in laboratory, due to the large splitting of the energy levels. However, one can easily identify from the mass-dependence of the discrete scaling factor that optimal situations, better than the equal-mass cases detected for example in [16], can occur for mass-imbalanced systems with $m_\alpha >> m_\beta$, when the level splitting could be more easily experimentally identified. For instance, in case of $m_\alpha = 100 m_\beta$, the ratio between consecutive levels of the bound-state energy spectrum is given by $\exp(2\pi/s_0) \sim 4.7$. Therefore, the actual experimental investigation are more promising in this direction, by considering the atom-molecule collision with two kind of atomic species [18, 30, 31]. By following this line of investigation in cold-atom experiments, we have the Heidelberg group studying the extreme mass-imbalance atomic mixtures composed of Caesium and Lithium atomic species [27].

With the actual possibilities to manipulate Lithium(Li)-Caesium(Cs) collisions [32], as well as considering that LiCs molecules can be generated in ultracold experiments [33], we understand that more favorable conditions are accessible by now to probe the rich Efimov physics in cold-atom laboratories by investigating the low-energy collision of a Cs atom in a weakly-bound LiCs molecule. In a more general perspective, by also including other possible atomic species, we can consider the scattering of an atomic specie α in a weakly-bound $\alpha-\beta$ system, for $m_\alpha >> m_\beta$. For that, in our following theoretical approach, we assume that the two-body interactions are such that there is no interaction between the two heavy particles. Further, by controlling the strength of the $\alpha - \beta$ binding in laboratory through Feshbach resonance mechanism, we understand that appropriate conditions can be created for atom-molecule collisions at very low energies.

Within our approach to identify the scaling behavior of scattering observables for three-body systems near the unitary limit, let us recall some recent discussion related to the $k \cot \delta_0^R$ pole behavior in the weakly bound three-body nuclear physics case of carbon-20 (^{20}C), within the three-body neutron-neutron$-^{18}$C model, where the subsystem neutron$-^{18}$C is bound. In this case, k is the on-energy shell momentum (incoming and final), with δ_0^R being the real part of the $s-$wave phase shift. This case, with two light particles and a heavy one, was shown to resemble the well-known equal-mass neutron-deuteron case, which was studied about forty years ago [34–36]. In contrast with these studies, when considering the present experimental possibilities in cold-atom laboratories, we can verify that the limit of two-heavy and one light particle is quite more interesting to be explored near the unitary limit. As in this limit, for the case $m_\alpha >> m_\beta$, we can identify several bound-state poles for the three-body spectrum with levels close together, in correspondence, a sequence of poles is expected to appear in the scattering observable $k \cot \delta_0^R$ for low-energy collisions of the heavy particle in the $\alpha - \beta$ bound

IOP Conf. Series: Journal of Physics: Conf. Series **915** (2017) 012002 doi:10.1088/1742-6596/915/1/012002

sub-system, which can be identified by the corresponding minima in the cross sections.

In the next, we have two sections reporting results for scattering observables in mass-asymmetric cases. Section 2 is concentrated in presenting some results and analysis related do the halo-nucleus ^{20}C, with section 3 having its focus in a more general mass-imbalanced atomic system with two heavy and one light particle. Section 4 gives some concluding remarks.

2. Universal properties in Neutron$-^{19}$C scattering

The low-energy properties of the elastic $s-$wave scattering for the $n-^{19}$C near the critical condition for the occurrence of an excited Efimov state in $n-n-^{18}$C was studied in Refs. [37,38], by considering renormalized zero-range as well as finite-range interactions. By fixing the two-neutrons separation energy in ^{20}C with available experimental data, it is studied the scaling of the real (δ_0^R) and imaginary parts of the $s-$wave phase-shift with the variation of the $n-^{18}$C binding energy. Universal characteristics are identified for the pole-position of $k\cot(\delta_0^R)$ and effective-range parameters. It was verified that the excited state of ^{20}C goes to a virtual state when increasing the $n-^{18}$C binding energy, resembling the neutron-deuteron behavior in the triton.

The basic formalism for a three-body halo nucleus formed by a core and two neutrons interacting via two-body separable potential is presented in Refs. [37,38]. The $n-n$ interaction was considered fixed, such that it was obtained the usual virtual-state energy, $E_{nn} = -143$ keV. For the $n-c$ subsystem ($E_{nc} = E_{^{19}\text{C}}$), the bound-state energies are varied within a range given by experimental available data, ranging from -160 ± 110 keV [39] to -530 ± 130 keV [40]. The $n-n-c$ three-body ground-state binding energy is fixed at $E_{^{20}\text{C}} = -\ 3.5$ MeV. The units for this case are such that $\hbar = 1$, with momentum variables in fm^{-1} and the unit conversion given by $\hbar^2/m_n = 41.47$ MeV fm^2. The core-mass number is defined as $A = m_c/m_n$, with $\mu_{nc} = Am_n/(A+1)$ being the reduced mass for the $n-c$ system, such that we have $A = 18$ and $\mu_{nc} = (18/19)m_n$ in the specific system we are considering.

The zero-range two-body interaction considered for the $s-$wave elastic $n-(nc)$ scattering formalism in Ref. [37] was extended to include finite-range two-body interactions, with separable form, given by

$$V_{ij}(p,p') = \lambda_{ij}\left(\frac{1}{p^2+\beta_{ij}^2}\right)\left(\frac{1}{p'^2+\beta_{ij}^2}\right), \tag{1}$$

where $ij = nn$ or nc, respectively, for the $n-n$ or $n-c$ two-body subsystems. λ_{ij} and β_{ij} refer to the strength and range r_{ij} of the respective two-body interaction. For negative bound or virtual two-body energies E_{ij}, the corresponding relations for the strengths and ranges are

$$\lambda_{ij} = \frac{-2\pi\mu_{ij}}{\beta_{ij}(\beta_{ij}\pm\kappa_{ij})}, \qquad r_{ij} = \frac{1}{\beta_{ij}} + \frac{2\beta_{ij}}{(\beta_{ij}\pm\kappa_{ij})^2}, \tag{2}$$

where $\kappa_{ij} = \sqrt{-2\mu_{ij}E_{ij}}$, with $(+)$ for bound and $(-)$ for virtual states.

By following the formalism given in Ref. [38] for the on-shell scattering amplitude $h_n(k;E)$, where E is the three-body energy $E \equiv E(k_i)$, we obtain:

$$h_n(k;E) = \frac{e^{\mathrm{i}\delta_0}\sin\delta_0}{k} = \frac{1}{k\cot\delta_0 - \mathrm{i}k}, \tag{3}$$

with the on-shell relative momentum given by $k \equiv k_i \equiv |\vec{k}_i| = |\vec{k}_f| = \sqrt{2\mu_{n(nc)}(E-E_{nc})}$ and the reduced mass is $\mu_{n(nc)} = m_n(A+1)/(A+2)$.

The validity of the picture found with the renormalized zero-range model in Ref. [37] is verified with finite-range interactions in [38] compatible with the nuclear force. The analysis follow a suggestion given in Ref. [34] for a parameterization of the effective-range expansion of $k\cot\delta_0^R$, where δ_0^R is the real part of the corresponding $s-$wave phase shifts. With E_0 being the pole position for the energy, we have

$$k\cot\delta_0^R = \frac{-a^{-1} + b\,E + c\,E^2}{1 - E/E_0}, \tag{4}$$

where a [$n-(nc)$ scattering length], b and c are the adjustable parameters. When considering short-ranged interactions, such that we have fixed the two-neutron separation energy in ^{20}C (3.5 MeV [39]) and the $n-n$ virtual state energy ($E_{nn} = -143$ keV), as the ^{19}C binding energy is varied from 0.2 to 0.8 MeV, the corresponding parameters are given in the following Table 1, where r_{nn} and r_{nc} are the corresponding ranges.

Table 1. One-neutron separation energy in ^{19}C, obtained by the given low-range parameters of the separable potential. The values of β_{nc} were obtained by fitting the two-neutron separation energy in ^{20}C (3.5 MeV [39]), with the $n-n$ interactions fixed by the virtual state energy, $E_{nn} = -143$ keV. In this case, we fix β_{nn} =24.5 fm^{-1} (r_{nn} =0.1228).

$\lvert E_{^{19}C}\rvert$(keV)	200	400	600	800	830
β_{nc}(fm^{-1})	18.970	17.036	15.592	14.395	14.234
r_{nc}(fm)	0.157	0.174	0.190	0.205	0.207

The parameters for the low-range potential shown in Table 1 provide results consistent with the ones obtained in Ref. [37] when using renormalized zero-range interactions. In Fig. 1 we present the results obtained by considering short-range interactions, reproduced from Ref. [38]. From this Fig. 1, by a fitting procedure, one can obtain the values for the parameters of the effective-range expansion (4). The corresponding parameters are shown in Table 2.

Table 2. Effective-range parameters, obtained by fitting Eq. (4) to Fig. 1, when considering different values of $\lvert E_{^{19}C}\rvert$ (first column) with short-range potentials.

$\lvert E_{^{19}C}\rvert$ (keV)	$-1/a$ (fm^{-1})	$10^4\,b$ (fm.keV)$^{-1}$	$10^8\,c$ (fm.keV^2)$^{-1}$	E_0 (keV)	d (fm^{-1})	r_{nc} (fm)
200	0.006	5.579	5.717	1304	0.831	0.157
400	0.066	6.742	9.144	749.0	0.622	0.174
600	0.234	9.840	13.16	402.9	0.652	0.190
800	1.798	44.67	35.19	78.86	2.153	0.205
830	5.149	119.8	85.78	28.98	5.497	0.207

Finally, we observe that in analogy with the known behavior of the elastic neutron-deuteron doublet $s-$wave scattering amplitude, the results presented in [37,38] confirm that the real part of the elastic $s-$wave phase shift (δ_0^R) reveals a zero when the $n-n-c$ system is close to an excited Efimov state (bound or virtual). For more details on this investigation, see Ref. [38], where the zero-range results are being compared with finite-range ones, by using separable one-term two-body interactions with different parametrizations. In this reference, results for the $s-$wave absorption parameter are also presented. By considering more realistic potential models, a recent investigation by Deltuva [41] is confirming the main conclusions on the universal properties verified for the $n-^{19}$C scattering in the models used in Refs. [37,38].

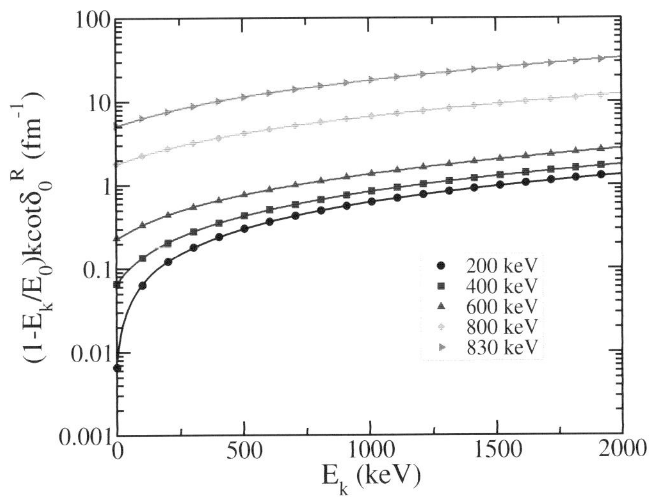

Figure 1. The function $(1 - E_K/E_0)k \cot \delta_0^R$, where $E \equiv E_K$ and E_0 is the pole position, is given in terms of $E_K \equiv k^2/(2\mu_{n,nc})$, by considering one-term separable low-range potentials (large β values), shown in Table 1. For each case, we plot the results obtained for different values of the two-body binding energy $|E_{nc}|$, as indicated inside the frame (adapted from Ref. [38]).

3. Scaling behavior in atom-dimer $\alpha - (\alpha - \beta)$ scattering with $m_\alpha \gg m_\beta$

As motivated in our introduction, large mass imbalanced systems can better be investigated in cold-atom experiments in view of possibilities to alter the two-body scattering length by tuning a Feshbach resonance with magnetic fields. Therefore, it is interesting to extend the approach considered in section 2 to the case of the collision of a heavy atom α on a dimer formed by α and a lighter atom β, in the case that $m_\alpha \gg m_\beta$. For this kind of system, one can also consider a Born-Oppenheimer (BO) approach in addition to the basic Faddeev three-body formalism that was used for the $n - (n -^{18}$ C) system, and check for the properties of the s-wave elastic cross section, while keeping the essence of the Efimov physics.

For that aim, we have examined the low-energy collision of a heavy particle α with a weakly-bound two-body $\alpha - \beta$ system. The heavy-heavy interaction is assumed zero as this should not affect the driving Efimov physics, which is originated by the heavy-light resonant interaction as already pointed out long ago by Fonseca and Shanley [7]. In this case the Efimov physics of the system is driven by the mass ratio and the ratio between the weakly-bound energy of the two-body system $B_{\alpha\beta}$ and a three-body energy scale, which can be chosen as the three-body bound state energy [25]. The two-body bound-state energy can also be represented by the corresponding very-large positive two-body scattering length, which is given by $a \equiv a_{\alpha\beta} = \sqrt{(2\mu_{\alpha\beta}/\hbar^2)/B_{\alpha\beta}}$, where $\mu_{\alpha\beta}$ is the reduced mass between α and β particles.

In this case, for a very large mass ratios $m_\alpha \gg m_\beta$, namely for our purpose we choose $m_\alpha/m_\beta = 20$ we can show that the minimum of the s-wave elastic cross-section appears and more than one. These minima are associated with the poles of $k \cot \delta_0^R$, which are associated with the long-range Efimov potential, strengthen by the mass ratio effect that implies a smaller geometrical ratio between the Efimov states as larger is the mass asymmetry in the heavy-heavy-light system.

Our results, which are verified for several small values of $B_{\alpha\beta}$, by using Faddeev approach as well as a BO approximation [42], are being exemplified in two panels of Fig. 2 for the cases that the mass-ratios is given by $A = m_\beta/m_\alpha = 0.08$ and 0.05, when considering $B_{\alpha\beta} = 0.05\mathcal{B}_3$, where $\mathcal{B}_3$ is an energy scale that we assume equal to the three-body ground-state energy. The novel feature emerging in these two panels is verified by the possibility that more zeros can be identified in the cross-section (or more poles in the $k\cot\delta_0^R$), as the mass asymmetry increases. This observation is closely related to the dominance of the Efimov long-range potential, which becomes more visible as the mass asymmetry increases. In a semi-classical description, such zeros should come as a consequence of the interference between different classical trajectories that pass around the two sides of the bound target. Therefore, it is conceivable that more zeros of the $s-$wave cross-section should appear as the intensity of the Efimov long-range potential raises.

As a final remark in this case, we should point that the ratio between the positions of the scattering momentum $k^{(n)}$ of the two-zeros shown in the upper plot of Fig. 2, for $A = 0.05$, is given by $\sqrt{E^{(n)}/E^{(n+1)}} = 6.5$, when the two-body energy is $B_{\alpha\beta} = 0.05\mathcal{B}_3$. This result, when compared with the corresponding Efimov ratio (note that $\sqrt{B_3^{(n)}/B_3^{(n+1)}} \approx 5$ for $B_{\alpha\beta} = 0$, in this mass asymmetric case), strongly suggests the plausibility of the above semi-classical description.

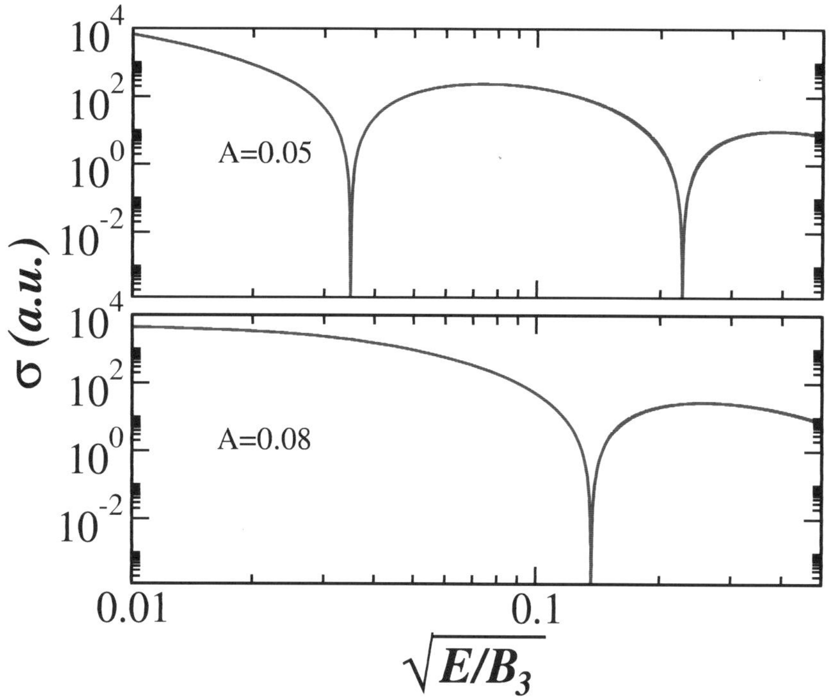

Figure 2. Results obtained for the total $s-$wave cross-section σ (in arbitrary units) as a function of $\sqrt{E/\mathcal{B}_3}$, by considering the mass-ratios $A \equiv m_\beta/m_\alpha$ =0.05 and 0.08, with $B_{\alpha\beta}/\mathcal{B}_3 = 0.05$. As shown, by decreasing the mass ratio A (from lower to upper panels), we can already identify the emergence of the second minimum in σ. For A =0.05, the ratio between the scattering momenta where the two minima are occurring (with $B_{\alpha\beta}/\mathcal{B}_3 = 0.05$), which is $\sim$ 6.5, can already be compared with the ratio between the Efimov levels in the three-body bound-state spectrum (where $\sqrt{B_3^{(n)}/B_3^{(n+1)}} \approx 5$ for $B_{\alpha\beta} = 0$).

4. Concluding remarks

In the introduction, our motivations to study mass-asymmetric systems in nuclear and atomic cases are presented. Following that, in section 2, we shortly review results already obtained on the low-energy properties of the elastic $s-$wave neutron$-{}^{19}$C elastic scattering near the critical condition for the occurrence of an excited Efimov state. The analysis of that results was done in close analogy with the case of the $s-$wave neutron-deuteron doublet phase-shift, where it is well known that at low-energies the properties of the neutron-deuteron doublet state is dominated by the Efimov physics, and the parameters of the effective range expansion of the neutron-deuteron doublet $s-$wave phase-shift present universal scaling laws with the triton binding energy for fixed nucleon-nucleon scattering lengths and effective ranges.

Next, by following the approach considered for the neutron$-{}^{19}$C scattering, it was communicated some examples on large mass-asymmetric cases of interest in recent cold-atom investigations. The low-energy s-wave elastic cross-section results were obtained for the collision between a heavy atom α and a weakly-bound heavy-light $(\alpha - \beta)$ sub-system. These studies are being motivated by the actual experimental possibilities, along the lines as reviewed in Ref. [27]. We should also point out the relevance of an extension of the present investigation to hetero-nuclear ultracold quantum gases.

As being verified, the scattering observable $k\cot\delta_0^R$ presents poles, corresponding to the zeros of the $s-$wave cross section, which can be directly connected with the Efimov physics, as observed when approaching the unitary limit. Here we have presented a sample of results for the mass-imbalanced system, considering $m_\beta/m_\alpha = 0.08$ and 0.05 in the case that $B_{\alpha\beta} = 0.05\mathcal{B}_3$. As shown, a second pole in $k\cot\delta_0^R$ can be found as we increase the imbalance between the masses of the components of the three-body system. A more detailed investigation is in progress.

Acknowledgments

This is a contribution in memory of Prof. S. A. Sofianos, a friend and active participant in the few-body community. For partial support, we thank the Brazilian agencies Coordenação de Aperfeiçoamento de Pessoal de Nível Superior - CAPES [Proc. no. 88881.030363/2013-01(MTY and MAS) and a Senior Visitor Program at the Instituto Tecnológico de Aeronáutica (LT)], Conselho Nacional de Desenvolvimento Científico e Tecnológico - CNPq [grants no. 302701/2013-3(MTY), 306191/2014-8(LT), 308486/2015-3 (TF)], and Fundação de Amparo á Pesquisa do Estado de São Paulo - FAPESP [grant no. 2016/01816-2(MTY)]. MRH acknowledges the partial support by National Science Foundation under Contract No. NSF-HRD-1436702 with Central State University and by the Institute of Nuclear and Particle Physics at Ohio University.

References

[1] Efimov V 1970 *Phys. Lett.* B **33** 563
[2] Amado R D and Noble J V 1971 *Phys. Lett.* B **35** 25; Amado R D and Noble J V 1972 *Phys. Rev.* D **5** 1992
[3] Adhikari S K, Delfino A, Frederico T, Goldman I D and Tomio L 1988 *Phys. Rev.* A **37** 3666
[4] Thomas L H 1935 *Phys. Rev.* **47**, 903
[5] Antunes A C, Baltar V L, Ferreira E M 1976 *Nucl. Phys.* A **265** 365
[6] Lim,T K, Duffy S K, Damer W C 1977 *Phys. Rev. Lett.* **38** 341
[7] Fonseca A, Redish E and Shanley P E1988 *Nucl. Phys.* A **320** 273
[8] Adhikari S K, Fonseca A C and Tomio L 1982 *Phys. Rev.* C **26** 77; Adhikari S K and Tomio L 1982 *Phys. Rev.* C **26** 83
[9] Cornelius Th and Glöckle W 1986 *J. Chem. Phys.* **85** 3906
[10] Dasgupta S, Mazumdar I, Bhasin V S 1994 *Phys. Rev.* C **50** R550
[11] Fedorov D V, Jensen A S, Riisager K 1994 *Phys. Rev. Lett.* **73** 2817
[12] Amorim A E A, Frederico T and Tomio L 1997 *Phys. Rev.* C **56** R2378
[13] Frederico T, Tomio L, Delfino A and Amorim A E A 1999 *Phys. Rev.* A **60** R9
[14] Kraemer T, Mark M, Waldburger P, Danzl J G, Chin C, Engeser B, Lange A D, Pilch K, Jaakkola A, Nägerl H-C and Grimm R 2006 *Nature* **440** 315

IOP Conf. Series: Journal of Physics: Conf. Series **915** (2017) 012002 doi:10.1088/1742-6596/915/1/012002

[15] Pollack S E, Dries D and Hulet R G 2009 *Science* **326** 1683
[16] Knoop S, Ferlaino F, Mark M, Berninger M, Schbel H, Ngerl H-C and Grimm R 2009 *Nature Physics* **5** 227
[17] Zaccanti M, Deissler B, D'Errico C, Fattori M, Jona-Lasinio M, Müller S, Roati G, Inguscio M and Modugno G 2009 *Nature Physics* **5** 586
[18] Barontini G, Weber C, Rabatti F, Catani J, Thalhammer G, Inguscio N and Minardi F 2009 *Phys. Rev. Lett.* **103** 043201; 2010 Erratum *Phys. Rev. Lett.* **104** 059901
[19] Inouye S *et al* 1998 *Nature* **392** 151
[20] Kunitski M *et al* 2015 *Science* **348** 551
[21] Nielsen E, Fedorov D, Jensen A and Garrido E 2001 *Phys. Rep.* **347** 373
[22] Jensen A S, Riisager K, Fedorov D V and Garrido E 2004 *Rev. Mod. Phys.* **76** 215
[23] Braaten E and Hammer H.-W. 2006 *Phys. Rep.* **428** 259
[24] Köhler T, Góral K, Julienne P S 2006 *Rev. Mod. Phys.* **78** 1311
[25] Frederico T, Delfino A, Tomio L and Yamashita M T 2012 *Prog. Part. Nucl. Phys.* **67** 939
[26] Ferlaino F and Grimm R 2010 *Physics* **3** 1
[27] Ulmanis J, Häfner S, Kuhnle E D and Weidemüller M 2016 *Natl. Sci. Rev.* **3** 174; Ulmanis J, Häfner S, Pires R, Kuhnle E D, Wang Y, Greene C H and Weidemüller M 2016 *Phys. Rev. Lett.* **117**153201; Pires R, Ulmanis J, Häfner S, Repp M, Arias A, Kuhnle E D and Weidemüller M 2014 *Phys. Rev. Lett.* **112** 250404
[28] Bhaduri R K, Chatterjee A, van Zyl B P 2011 *Am. J. Phys.* **79** 274; 2012 *Am. J. Phys.* **80** 94
[29] Coon S A and Holstein B R 2002 *Am. J. Phys.* **70** 513
[30] Bloom R S, Hu M-G, Cumby T D and Jin D S 2013 *Phys. Rev. Lett.* **111** 105301
[31] Hu M G, Bloom R S, Jin D S, Goldwin J M 2014 *Phys. Rev.* A **90** 013619
[32] Krems R V 2006 *Phys. Rev. Lett.* **96** 123202
[33] Kraft S D, Staanum P, Lange J, Vogel L, Wester R and Weidemüller M 2006 **39** S993
[34] van Oers W T H and Seagrave J D 1967 *Phys. Lett.* B **24** 562
[35] Whiting J S and Fuda M G 1976 *Phys. Rev.* C **14** 18
[36] Girard B A and Fuda M G 1979 *Phys. Rev.* C **19** 579
[37] Yamashita M T, Frederico T and Tomio L 2008 *Phys. Lett.* B **660** 339; 2008 *Phys. Lett.* B **670** 49
[38] Shalchi M A, Yamashita M T, Hadizadeh M R, Frederico T and Tomio L 2016 *Phys. Lett.* B **764** 196; 2017 Erratum *Phys. Lett.* B https://doi.org/10.1016/j.physletb.2017.05.070
[39] Audi G, Wapstra A H and Thibault C 2003 *Nucl. Phys.* A **729** 337
[40] Nakamura T *et al* 1999 *Phys. Rev. Lett.* **83** 1112
[41] Deltuva A 2017 *Phys. Lett.* B https://doi.org/10.1016/j.physletb.2017.07.036
[42] Shalchi M A, Yamashita M T, Hadizadeh M R, Garrido E, Tomio L and Frederico T 2017 Probing the Efimov discrete scaling in atom-molecule collision *Preprint* arXiv:1708.00034

Scattering length calculations via Faddeev approach

E A Kolganova

Joint Institute for Nuclear Research, Dubna, 141980, Russia

E-mail: kea@theor.jinr.ru

Abstract. The results of calculation of the scattering length for the collision of a ^{4}He atom with a ^{4}He^{7}Li dimer are presented. These investigations are based on the hard-core version of the Faddeev differential equations.

1. Introduction

Weakly bound small atomic clusters attracted considerable attention in recent years especially due to the possibility to observe states of Efimov nature [1]. Interaction between neutral atoms is of the short-range van-der-Waals type and in some cases can almost or just barely support a weakly two-body bound state. In such a cases the scattering length will be much larger than the effective range of the interaction. Efimov discovered that in a three-body system an effective long-range attraction arises, and this attraction may support an infinite number of bound states. The binding energy of these states is arranged in a geometric series with an accumulation point at zero energy threshold. The energies of the Efimov levels are universally related and this relation does not depend on the form of the pair-wise interactions in the three-body system [2].

One of a promising atomic species for observing Efimov states is ^{4}He three-atomic cluster. The interaction between two helium atoms is quite small and supports only one bound state with the energy about 1mK and so a very large scattering length around 100 Å. An excited state of Efimov nature in helium trimer was theoretically predicted (see, [3, 4] and refs. therein). Most recently this has been brilliantly confirmed experimentally using the combination of Coulomb explosion imaging with cluster mass selection by matter wave diffraction [5].

The implementation of magnetic Feshbach resonances in ultracold atoms experiments gives a possibility to change the scattering length by simply varying the strength of an applied magnetic field. The first observation of Efimov-type resonance in an ultracold gas of cesium atoms has been reported in [6]. Experimentally, in [6] Efimov resonance was observed as a giant three-body recombination loss when the intensity of the magnetic field and so the strength of the two-body interaction was varied. Striking manifestations of the Efimov effect have been predicted for three-body recombination processes in ultracold gases with tunable two-body interactions in [7, 8]. Although in this experiment only one Efimov resonance was observed, recently the second Efimov level has been measured using the same technique [9]. Now it has become possible to study Efimov effect in many other atomic systems [10, 11, 12, 13, 14, 15]. Review of the latest theoretical and experimental advances in Efimov physics can be found in [16].

There is a growing interest in the study of He_2 - Alkali van-der-Waals systems, which are expected to be of Efimov nature. As in case of a ^{4}He dimer, the He - alkali-atom interactions also support a single

IOP Conf. Series: Journal of Physics: Conf. Series **915** (2017) 012003 doi:10.1088/1742-6596/915/1/012003

weakly bound state. So, in triatomic $^4He_2Li$ system one can expect the existence of Efimov levels and a large He - HeLi scattering length which is investigated in this work.

2. Method

In describing a three-atomic system we use the standard reduced Jacobi coordinates $\boldsymbol{x}_\alpha, \boldsymbol{y}_\alpha$, $\alpha = 1,2,3$, expressed in terms of the position vectors of the atoms $\boldsymbol{r}_\alpha$ and their masses m_α [17]:

$$\begin{aligned}\boldsymbol{x}_\alpha &= \left[\frac{2\mathrm{m}_\beta \mathrm{m}_\gamma}{\mathrm{m}_\beta+\mathrm{m}_\gamma}\right]^{1/2} (\boldsymbol{r}_\beta - \boldsymbol{r}_\gamma)\\ \boldsymbol{y}_\alpha &= \left[\frac{2\mathrm{m}_\alpha(\mathrm{m}_\beta+\mathrm{m}_\gamma)}{\mathrm{m}_\alpha+\mathrm{m}_\beta+\mathrm{m}_\gamma}\right]^{1/2}\left(\boldsymbol{r}_\alpha - \frac{\mathrm{m}_\beta \boldsymbol{r}_\beta + \mathrm{m}_\gamma \boldsymbol{r}_\gamma}{\mathrm{m}_\beta+\mathrm{m}_\gamma}\right)\end{aligned} \tag{1}$$

where (α,β,γ) stand for a cyclic permutation of the atom numbers $(1,2,3)$. The coordinates $\boldsymbol{x}_\alpha, \boldsymbol{y}_\alpha$ determine the six-dimensional vector $X \equiv (\boldsymbol{x}_\alpha, \boldsymbol{y}_\alpha)$. Relations between them are given by orthogonal transform Jacobi vectors with different α

$$\boldsymbol{x}_\beta = \mathrm{c}_{\beta\alpha}\boldsymbol{x}_\alpha + \mathrm{s}_{\beta\alpha}\boldsymbol{y}_\alpha, \qquad \boldsymbol{y}_\beta = -\mathrm{s}_{\beta\alpha}\boldsymbol{x}_\alpha + \mathrm{c}_{\beta\alpha}\boldsymbol{y}_\alpha, \tag{2}$$

where

$$\begin{aligned}\mathrm{c}_{\alpha\beta} &= -\left(\frac{\mathrm{m}_\alpha \mathrm{m}_\beta}{(\mathrm{m}_\alpha+\mathrm{m}_\beta)(\mathrm{m}_\beta+\mathrm{m}_\gamma)}\right)^{1/2},\\ \mathrm{s}_{\alpha\beta} &= (-1)^{\beta-\alpha}\,\mathrm{sign}(\beta-\alpha)\left(1-\mathrm{c}^2_{\alpha\beta}\right)^{1/2}.\end{aligned}$$

In the following the Helium atoms are assigned the numbers 1 and 3 while the Lithium atom has the number 2. Since two atoms of ^{4}He are identical bosons the corresponding Faddeev component $F_2(\boldsymbol{x}_2, \boldsymbol{y}_2)$ is invariant under the permutation of the 1 and 3 particles

$$F_2(-\boldsymbol{x}_2, \boldsymbol{y}_2) = F_2(\boldsymbol{x}_2, \boldsymbol{y}_2). \tag{3}$$

The identity of the two ^{4}He atoms also implies that the Faddeev components $F_1(\boldsymbol{x}_1, \boldsymbol{y}_1)$ and $F_3(\boldsymbol{x}_3, \boldsymbol{y}_3)$ are obtained from each other by a simple rotation of the coordinate space. Thus, we only have two independent Faddeev components: $F_2(\boldsymbol{x}, \boldsymbol{y})$, which is associated with the ^{4}He–^{4}He subsystem, and $F_1(\boldsymbol{x}, \boldsymbol{y})$ - associated with a pair of Li and ^{4}He atoms. The resulting hard-core Faddeev equations read

$$(-\Delta_X - E)F_\alpha(\boldsymbol{x}_\alpha, \boldsymbol{y}_\alpha) = \begin{cases} 0, & |\boldsymbol{x}_\alpha| < c \\ -V_\alpha(\boldsymbol{x}_\alpha)\Psi^{(\alpha)}(\boldsymbol{x}_\alpha, \boldsymbol{y}_\alpha), & |\boldsymbol{x}_\alpha| > c \end{cases}, \tag{4}$$

$$\Psi^{(\alpha)}(\boldsymbol{x}_\alpha, \boldsymbol{y}_\alpha)\Big|_{|\boldsymbol{x}_\alpha|=c} = 0, \tag{5}$$

$$\alpha = 1,2,$$

where $\Psi^{(1)}$ and $\Psi^{(2)}$ denote the total wave function $\Psi(X)$ of the Li^4He_2-system written in terms of the Faddeev components F_1 and F_2 in coordinates $\boldsymbol{x}_1, \boldsymbol{y}_1$ and $\boldsymbol{x}_2, \boldsymbol{y}_2$, respectively. By c we denote the hard-core radius. This radius was taken the same (in coordinates $\boldsymbol{x}_\alpha$) for all three inter-atomic interaction potentials and was chosen in such a way that any further decrease of it does not affect the trimer ground-state energy. A detailed description of the Faddeev differential equations in the hard-core model in case of symmetric helium trimer can be found in [18]. By V_1 we denote the interatomic Li–He potential and V_2 - the He–He potential adjusted to the corresponding reduced Jacobi coordinates $\boldsymbol{x}_1$ and $\boldsymbol{x}_2$, respectively.

IOP Conf. Series: Journal of Physics: Conf. Series **915** (2017) 012003 doi:10.1088/1742-6596/915/1/012003

In the present investigation we apply the above formalism to the ^{7}Li^{4}He$_2$ three-atomic systems with total angular momentum $L=0$. Expanding the functions F_1 and F_2 in a series of bispherical harmonics we have

$$F_\alpha(\mathbf{x},\mathbf{y}) = \sum_l \frac{f_l^{(\alpha)}(x,y)}{xy}\mathscr{Y}_{ll0}(\widehat{x},\widehat{y}), \quad \alpha=1,2, \tag{6}$$

where $x=|\mathbf{x}|$, $y=|\mathbf{y}|$, $\widehat{x}=\mathbf{x}/x$, and $\widehat{y}=\mathbf{y}/y$. As a result the equations (4) and boundary conditions (5) are transformed to the following partial integro-differential equations

$$\left(-\frac{\partial^2}{\partial x^2}-\frac{\partial^2}{\partial y^2}+l(l+1)\left(\frac{1}{x^2}+\frac{1}{y^2}\right)-E\right)f_l^{(\alpha)}(x,y)$$
$$=\left\{\begin{array}{ll} 0, & x<c \\ -V_\alpha(x)\psi_l^{(\alpha)}(x,y), & x>c \end{array}\right., \qquad \alpha=1,2, \tag{7}$$

and partial boundary conditions

$$\psi_l^{(\alpha)}(x,y)\Big|_{x=c}=0, \qquad \alpha=1,2. \tag{8}$$

The partial wave functions $\psi_l^{(\alpha)}$, $\alpha=1,2$, read as follows

$$\psi_l^{(\alpha)}(x,y) = f_l^{(\alpha)}(x,y)+\sum_{l',\beta\neq\alpha}\int_0^1 d\eta\, h^0_{(\alpha;ll0)(\beta;l'l'0)}(x,y,\eta)f_{l'}^{(\beta)}(x_{\beta\alpha}(\eta),y_{\beta\alpha}(\eta))$$

where (cf. [17])

$$\begin{aligned}
&h^L_{(\alpha;l\lambda L)(\beta;l'\lambda' L)}(x,y,\eta) = \\
&\frac{xy}{x_{\beta\alpha}(\eta)y_{\beta\alpha}(\eta)}(-1)^{l+L}\frac{(2\lambda+1)(2l+1)}{2^{\lambda+l}}\left[(2\lambda)!(2l)!(2\lambda'+1)(2l'+1)\right]^{1/2} \\
&\times\sum_{k=0}^{k_{max}}(-1)^k(2k+1)P_k(\eta)\sum_{\substack{\lambda_1+\lambda_2=\lambda,\\ l_1+l_2=l}}\frac{y^{\lambda_1+l_1}x^{\lambda_2+l_2}}{[y_{\beta\alpha}(\eta)]^\lambda[x_{\beta\alpha}(\eta)]^l}(-1)^{\lambda_1}\mathrm{c}_{\beta\alpha}{}^{\lambda_1+l_2}\mathrm{s}_{\beta\alpha}{}^{\lambda_2+l_1} \\
&\times\left[(2\lambda_1)!(2l_1)!(2\lambda_2)!(2l_2)!\right]^{-1/2}\sum_{\lambda''l''}(2\lambda''+1)(2l''+1)\begin{pmatrix}\lambda_1 & l_1 & \lambda''\\ 0&0&0\end{pmatrix} \\
&\times\begin{pmatrix}\lambda_2 & l_2 & l''\\ 0&0&0\end{pmatrix}\begin{pmatrix}k & \lambda'' & \lambda'\\ 0&0&0\end{pmatrix}\begin{pmatrix}k & l'' & l'\\ 0&0&0\end{pmatrix} \\
&\times\left\{\begin{array}{ccc} l' & \lambda' & L\\ \lambda'' & l'' & k\end{array}\right\}\left\{\begin{array}{ccc}\lambda_1 & \lambda_2 & \lambda\\ l_1 & l_2 & l\\ \lambda'' & l'' & L\end{array}\right\}, \\
&k_{max}=\frac{1}{2}(l+\lambda+l'+\lambda').
\end{aligned} \tag{9}$$

Here $P_k(\eta)$ is the Legendre polynomial of order k. The standard notation for the 3-j, 6-j, and 9-j Wigner symbols is used. We also use the notation

$$x_{\beta\alpha}(\eta) = \sqrt{\mathrm{c}^2_{\beta\alpha}x^2+2\mathrm{c}_{\beta\alpha}\mathrm{s}_{\beta\alpha}xy\eta+\mathrm{s}^2_{\beta\alpha}y},$$
$$y_{\beta\alpha}(\eta) = \sqrt{\mathrm{s}^2_{\beta\alpha}x-2\mathrm{c}_{\beta\alpha}\mathrm{s}_{\beta\alpha}xy\eta+\mathrm{c}^2_{\beta\alpha}y}.$$

The asymptotic boundary condition for the partial-wave Faddeev components of the scattering wave function for $\rho = \sqrt{x^2+y^2} \to \infty$ and/or $y \to \infty$ reads as follows (cf. [20])

$$f_l^{(\alpha)}(x,y,p) = \delta_{l0}\psi_d^{(\alpha)}(x)\left\{sin(py)+\exp(ipy)\left[a_0^{(\alpha)}(p)+o\left(y^{-1/2}\right)\right]\right\} + \frac{\exp(i\sqrt{E}\rho)}{\sqrt{\rho}}\left[A_l^{(\alpha)}(E,\theta)+o\left(\rho^{-1/2}\right)\right] \quad (10)$$

where $p = \sqrt{E-\varepsilon_d^{(\alpha)}}$ is the momentum conjugate to the coordinate y and E is the scattering energy. $\varepsilon_d^{(\alpha)}$ stands for the correspondent dimer energy while $\psi_d^{(\alpha)}(x)$ denotes the dimer wave function which is assumed to be zero within the core, that is, $\psi_d^{(\alpha)}(x) \equiv 0$ for $x \le c$. The coefficient $a_0^{(\alpha)}(p)$ is the elastic scattering amplitude and $A_l^{(\alpha)}(E,\theta)$ are the corresponding partial-wave Faddeev breakup amplitudes. The $^4He - {}^4He^7Li$ scattering length ℓ_{sc} is given by

$$\ell_{sc} = -\sqrt{\frac{m_1+m_2+m_3}{2m_1(m_2+m_3)}}\lim_{p\to 0}\frac{a_0^{(1)}(p)}{p}. \quad (11)$$

Here we only deal with a finite number of equations (7), assuming $l \le l_{max}$ where l_{max} is a certain non-negative integer. As in [18, 19] we use a finite-difference approximation of the boundary-value problem (7), (8), (10) in the polar coordinates ρ and θ. The grids are chosen such that the points of intersection of the arcs $\rho = \rho_i$, $i = 1,2,\ldots,N_\rho$ and the rays $\theta = \theta_j$, $j = 1,2,\ldots,N_\theta$ with the core boundary $x = c$ constitute the knots. The value of the core radius is chosen to be $c = 1$ Å by the argument given in [19]. We also follow the same method for choosing the grid radii ρ_i (and, thus, the grid hyperangles θ_j) as described in [18, 19] in details. Atomic massed of the isotopes are taken from [21].

3. Results and Discussion

Our calculations are based on the semi-empirical LM2M2 [22] potential by Aziz and Slaman for He-He interaction, and the theoretically derived KTTY [23] potential by Kleinekathöfer, Tang, Toennies and Yiu for Li-He interaction with more accurate coefficients taken from [24]. Both of these potentials are widely used in the literature and presented in Fig. 1 as well as all other He - Alkali atom KTTY potentials from [24]. It is clearly seen that He-Alkali atom potentials become much shallower and wider with increasing mass of alkali atom. Calculated values of the binding energy for correspondent dimers are presented in Table 1. All He-Alkali dimers are weakly bound and in the case of HeLi and HeCs systems the binding energy is of the same order as in helium dimer. It gives indication on possible existence of Efimov states in correspondent He_2 - Alkali atom triatomic systems. Indeed, in calculations [25, 26, 27, 28, 29] the excited state with the energy very close to the LiHe threshold has been found. However, different methods demonstrate a large discrepancy between the results (see Table 1 [28] or Table II [29]). The energy levels of He_2Li trimer are found to be very sensitive to the details of the numerical methods and the interaction potentials. We can conclude that this triatomic system is much more complicated for investigation than the helium trimer studied previously [3, 4].

The scattering length calculations are very sensitive to the grid parameters. To investigate this sensitivity we calculate the scattering length with increasing cutoff hyperradius ρ_{max} up to 500 Å and increasing grid numbers $N \equiv N_\rho = N_\theta$ up to 2500. The value of l_{max} has been increased up to 4. The results for the ^4He -$^4He^7Li$ scattering length ℓ_{sc} as a function of the grid dimension N are presented in Fig.2. Convergence of the scattering length value is essentially achieved for $N \sim 2500$. Increasing values of the cutoff hyperradius ρ_{max} from 250 Å to 500 Å change the value of ℓ_{sc} by 2%. Taking into account partial angular momentum $l_{max} = 2$ is increase the scattering length by 10% while $l_{max} = 4$ gives only 2% . As a result we found that the ^4He -$^4He^7Li$ scattering length for the LM2M2 and KTTY potentials is negative and equals -267 Å for the largest grid that we could achieve. The large value of the scattering

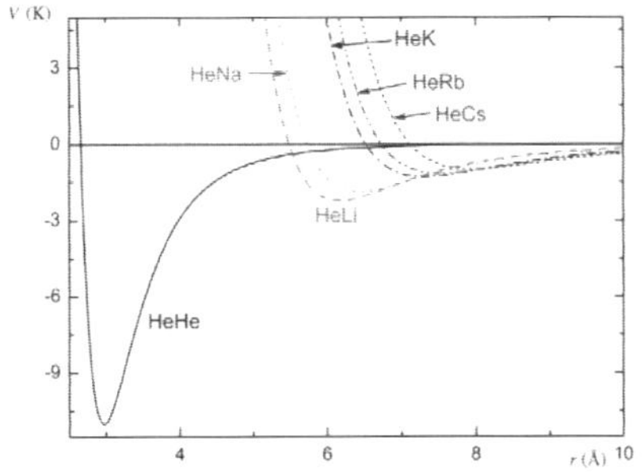

Figure 1. The He–He LM2M2 potential and He - Alkali atoms KTTY potentials V (in K) as a function of the interatomic distance r (in Å).

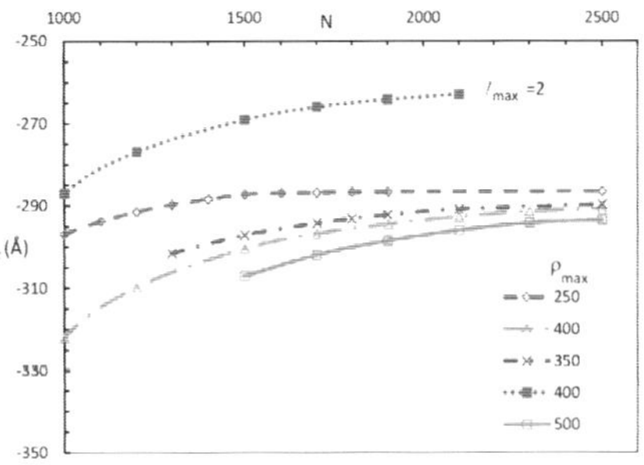

Figure 2. The ^{4}He -^{4}He^{7}Li scattering length ℓ_{sc}(Å) as a function of the grid dimension $N \equiv N_\rho = N_\theta$. The curves presented are distinguished by the value of the cutoff hyperradius. The upper curve shows the results obtained for $l_{max} = 2$, other curves correspond to $l_{max} = 0$.

Table 1. Absolute values of dimer energies $|\varepsilon_d|$ (in mK) for He-He and ^{4}He–Alkali atom for the potentials used.

Dimer	$\vert\varepsilon_d\vert$ (mK)	Dimer	$\vert\varepsilon_d\vert$ (mK)
^{4}He^{3}He	–	^{4}He^{23}Na	28.97
^{4}He^{4}He	1.310	^{4}He^{39}K	11.20
^{4}He^{6}Li	1.512	^{4}He^{85}Rb	10.27
^{4}He^{7}Li	5.617	^{4}He^{133}Cs	4.945

length supports the assumption that ^{4}He$_2$^{7}Li is the system of Efimov type. The negative sign means that this system is more likely to have the virtual state than the excited state. However, it could also mean that asymptotic behavior at used cutoff hyperradius $\sim$ 500 Å has not yet started. Studying of this fact requires further investigation.

In conclusion, the results of calculation of the scattering length for the collision of a ^{4}He atom on a ^{4}He^{7}Li dimer are presented. Its large value indicate that ^{4}He$_2$^{7}Li system is of Efimov type.

Acknowledgments

The author thanks E. Hiyama, P. Stipanović, A. Kievsky, E. Yarevsky, A.K. Motovilov and R.V. Jolos for valuable discussions.

[1] Efimov V 1970 *Phys. Lett.* B **33** 563
[2] Efimov V N 1970 *Sov. J. Nucl. Phys.* **12** 589
[3] Kolganova E A, Motovilov A K and Sandhas W 2011 *Few-Body Syst.* **51** 249
[4] Kolganova E A, Motovilov A K and Sandhas W 2017 *Few-Body Syst.* **58** 35
[5] Kunitski M, Zeller S, Voigtsberger J, Kalinin A, Schmidt L Ph H, Schöffer M, Czasch A, Schöllkopf W, Grisenti R E, Janke C, Blume D and Dörner R 2015 *Science* **348** 551
[6] Kraemer T, Mark M, Waldburger P, Danzl J G, Chin C, Engeser B, Lange A D, Pilch K, Jaakkola A, Nağerl H-C and Grimm R 2006 *Nature* **440** 315
[7] Esry B D, Greene C H and Burke J P 1999 *Phys. Rev. Lett.* **83** 1751
[8] Braaten E and Hammer H-W 2006 *Phys. Rep.* **428** 259
[9] Huang B, Sidorenkov L A and Grimm R 2014 *Phys. Rev. Lett.* **112** 190401

[10] Barontini G, Weber C, Rabatti F, Catani J, Thalhammer G, Inguscio M and Minardi F 2009 *Phys. Rev. Lett.* **103** 043201; Erratum 2010 **104** 059901
[11] Lompe T, Ottenstein T B, Serwane F, Viering K, Wenz A N, Zürn G and Jochim S 2010 *Science* **330** 940
[12] Williams J R, Hazlett E L , Huckans J H, Stites R W, Zhang Y and O'hara K M 2009 *Phys. Rev. Lett.* **103** 130404
[13] Rem B S, Grier A T, Ferrier-Barbut I, Eismann U, Langen T, Navon N, Khaykovich L, Werner F, Petrov D S, Chevy F and Salomon C 2013 *Phys. Rev. Lett.* **110** 163202
[14] Bloom R S, Hu M-G, Cumby T D and Jin D S 2013 *Phys. Rev. Lett.* **111** 105301
[15] Ulmanis J, Häfner S, Pires R, Werner F, Petrov D S, Kuhnle E D and Weidemüller M 2016 *Phys. Rev.* A **93** 022707
[16] Naidon P and Endo Sh 2017 *Rep. Prog. Phys.* **80** 056001
[17] Faddeev L D and Merkuriev S P 1993 *Quantum scattering theory for several particle systems* (Kluwer Academic Publishers, Doderecht)
[18] Kolganova E A , Motovilov A K and Sofianos S A 1998 *J. Phys.* B **31** 1279
[19] Motovilov A K, Sandhas W, Sofianos S A and Kolganova E A 2001 *Eur. Phys. J.* D **13** 33
[20] Kolganova E A , Motovilov A K and Sofianos S A 2004 *Phys. Rev.* A **80** 052711
[21] Mills I, Cvitaš T, Homann K, Kallay N and Kuchitsu K 1993 *Quantities, Units and Symbols in Physical Chemistry* (2nd edition, Blackwell Science, Oxford)
[22] Aziz R A and Slaman M J 1991 *J. Chem. Phys.* **94** 8047
[23] Kleinekathöfer U, Tang K T, Toennies J P and Yiu C I 1996 *Chem. Phys. Lett.* **249** 257
[24] Kleinekathöfer U, Lewerenz M and Mladenoc M 1999 *Phys. Rev. Lett.* **83** 4717
[25] Suno H and Esry B D 2009 *Phys. Rev.* A **80** 062702
[26] Suno H, Hiyama E and Kamimura M 2013 *Few-Body Syst.* **54** 1557
[27] Wu M-S, Han H-L, Li Ch-B and Shi T-Y 2014 *Phys. Rev.* A **90** 062506
[28] Kolganova E A 2017 *Few-Body Syst.* **58** 57
[29] Suno H 2017 *Phys. Rev.* A **96** 012508

IOP Conf. Series: Journal of Physics: Conf. Series **915** (2017) 012004 doi:10.1088/1742-6596/915/1/012004

Construction of lambda-nucleon s-wave potential through quantum inverse scattering at fixed angular momentum

E F Meoto and M L Lekala

Department of Physics, University of South Africa, P. O. Box 392, 0003 Pretoria, South Africa

E-mail: meotoef@unisa.ac.za or emilemeoto@aims.ac.za

Abstract. Quantum systems with a strangeness degree of freedom are very important as they provide an extra dimension, and hence a deeper insight into nuclear matter. Usually phenomenological potentials obtained through meson exchange theories are used in investigating these hypernuclear systems. In this paper potentials for lambda-nucleon interactions in the spin singlet and spin triplet states, constructed through fixed-angular momentum inversion based on Marchenko theory, are presented. Owing to experimental difficulties in obtaining a sufficient number of lambda-nucleon scattering events, theoretical phase shifts are used as input for the inversion. The constructed potential is energy-independent, making it more suitable for quantum-mechanical few-body calculations.

1. Introduction

The choice of a suitable baryon-baryon potential for use in simulations has been one of the preoccupations right at the heart of nuclear physics for much of its history. In particular, the last two or three decades have seen an increased effort towards understanding baryon-baryon interactions for systems with a strangeness degree of freedom. Much of the effort has been directed towards single and double strangeness systems. These potentials are important for many reasons. For example, they are used as input in simulations of hypernuclei. The glue-like role of hyperons within hypernuclei and the non-mesonic decay of hyperons in large-baryon number systems through the weak interaction are phenomena requiring further elucidation, that may only come from more reliable hyperon-nucleon potentials. These interactions also provide a deeper understanding of multistrangeness systems such as present at the core of a neutron star, whose equation of state depends on hyperon-nucleon and hyperon-hyperon potentials.

Meson-exchange theories have been the principal framework for developing most of the hyperon-nucleon and hyperon-hyperon potentials currently in use for few-body calculations. The most widely used meson-exchange potentials are the soft-core versions of the Nijmegen potentials [1], which were first formulated in [2]. A historical review of the development of the Nijmegen potentials is found in [3] and [4]. While most of these potentials do successfully address certain important issues such as the overbinding problem, they still suffer some inadequacies. The sensitivity of few-body calculations to the choice of potentials is symptomatic for these inadequacies, within the error limits of the few-body method used. Could these inadequacies stem from a much wider problem of nuclear physics using potentials constructed between structureless hadrons, even at energy scales for which the quark degrees of freedom in Quantum Chromodynamics (QCD) represent the "correct" physics? Already, there has been considerable effort in building new potentials based on QCD. The Kioto-Niigata potential for all interactions between spin-1/2 octet baryons (n, p, Σ^+, Σ^0, Σ^-, Ξ^-, Ξ^0 and Λ) is one such development [5–7]. However, these quark-model potentials have not enjoyed as widespread use in few-body calculations as the meson-exchange potentials. A more systematic inclusion of QCD theories in nuclear physics *may* hold more for the future of nuclear physics, as QCD is the underlying theory of the strong interaction. The ambiguities of the currently used hypernuclear potentials make it necessary to investigate potentials obtained using a different theory.

In this paper we propose new potentials for the ΛN interaction in the spin singlet and triplet states, constructed not using QCD theories, but through quantum inverse scattering theory. The fixed-angular momentum approach has been used in solving this inverse scattering problem. Owing to its simplicity when compared to Gel'fand-Levitan inversion, the Marchenko inversion scheme is used. The rest of the paper is organised as follows: Section 2 provides a brief recapitulation of quantum inverse scattering at fixed angular momentum, Section 3 is devoted to ΛN scattering experiments and theoretical scattering data, while Sections 4 and 5 contain the results and conclusion, respectively.

2. The scattering matrix to potential inverse problem

2.1. Marchenko theory

Inversion may either be performed from experimental observables to the potential, or from the scattering matrix to the potential. The inversion from scattering matrix to potential may either be done at a fixed angular momentum (ℓ) or at a fixed energy (k), or using a hybrid formalism for a given energy range and set of angular momenta. The normalisation $\hbar^2 = 2\mu = 1$ is used. A detailed review on inverse scattering theory is found in the monograph by [8], while an up-to-date review on applications in nuclear physics may be found in [9]. Of interest in this paper is Marchenko inversion, one of the fixed-angular momentum formalisms. For a short-range potential $V_\ell(r)$ with spherical symmetry the potential at each partial wave ℓ is constructed by solving the following Fredholm integral equation for $K_\ell(r,r')$:

$$K_\ell(r,r') + A_\ell(r,r') + \int_r^\infty K_\ell(r,s)A_\ell(s,r')ds = 0 \tag{1}$$

Here $A_\ell(r,r')$ is a symmetric input kernel. This input kernel is constructed as in [10], taking into account any bound states:

$$A_\ell(r,r') = \frac{1}{2\pi}\int_{-\infty}^{\infty} h_\ell^+(k,r)\left\{1 - S_\ell(k)\right\} h_\ell^+(k,r')dk + \sum_{i=1}^{n_b} M_i h_\ell^+(k_i,r)h_\ell^+(k_i,r') \tag{2}$$

where $S_\ell(k)$ is the scattering matrix, n_b is the number of physical bound states, if there are any, and $h_\ell^+(k,r)$ are Riccati-Hankel functions. The M_i are normalisation constants for the bound-state wavefunctions. The unphysical bound states, also known as the Pauli forbidden states, which may arise are usually removed through supersymmetry. The functions $h_\ell^+(k,r)$ are defined as follows:

$$h_\ell^+(k,r) = ie^{i\pi\ell}\left(\frac{\pi kr}{2}\right)^{\frac{1}{2}} H^{(1)}_{\ell+\frac{1}{2}}(k,r) \tag{3}$$

IOP Conf. Series: Journal of Physics: Conf. Series **915** (2017) 012004 doi:10.1088/1742-6596/915/1/012004

where $H_n^{(1)}(k,r)$ is a Hankel function of the first kind of order n. For the case of interest, $\ell = 0$, the Riccati-Hankel functions take the form

$$h_0^+(k,r) = i\left(\frac{\pi kr}{2}\right)^{\frac{1}{2}} H_{\frac{1}{2}}^{(1)}(k,r) \tag{4}$$

In inversion theory $K_\ell(r,r')$ arises as the kernel of a transformation from the solution of the radial Schrödinger equation for a free particle to any solution of the radial Schrödinger equation, for example the Jost solutions [11, 12]. Furthermore, this output kernel, which has the property of strict upper triangularity ($K_l(r,r') = 0, r > r'$), satisfies a Goursat problem which is right at the heart of quantum inverse scattering [13–18]. The auxiliary condition of this Goursat problem on the characteristic curve $r = r'$ ensures that the derivative of the diagonal entries in the kernel provides the scattering potential i.e.

$$-2\frac{d}{dr}K_\ell(r,r) = V_\ell(r) \tag{5}$$

Therefore, solving the inverse problem of scattering matrix to potential can be understood in proper mathematical terms as reconstructing the auxiliary conditions of this Goursat problem.

Marchenko inversion theory has been used to construct particle-particle potentials, in particular nucleon-nucleon potentials [19–23]. [9] contains an extensive list of references for particle-cluster and cluster-cluster potentials that have been constructed through inversion theory, for example the neutron-alpha potential.

Here, we apply Marchenko inversion theory to construct ΛN potentials. A motivating factor for such an attempt stems from the fact that the widely used models for the ΛN potential disagree with experimental results in terms of single lambda binding energies (B_Λ) of ground state and excited states, even for low-baryon number hypernuclei.

2.2. Regularisation of linear system

Theory requires that $S_\ell(k)$, which is a complex-valued function, should be known for all energies or wavenumbers i.e. $k \in [0, \infty)$. However, the scattering matrix is obtained from the phase shift, which is itself derived from experimental observables: $S_\ell(k) = \exp(2i\delta_\ell(k))$, where $\delta_\ell(k)$ is the phase shift. The energy range of these experimental observables are limited by laboratory circumstances, thus limiting the energy range for which $S_\ell(k)$ is known. With such a scattering matrix, the inverse problem of computing the potential from the scattering matrix results in a linear system in which there are more degrees of freedom than there are constraints to restrict these degrees of freedom.

In order to solve this inverse problem, some form of regularisation is needed, so as to provide additional information. The most common approach, which is also the one that has been used in this paper, is to interpolate $S_\ell(k)$ using a rational function, and then extrapolating it to sufficiently high energies. Making use of the Bargmann rational-function representations of the Jost functions [8], the following interpolation, which explicitly shows the singularities of the Jost functions, is used:

$$S_\ell(k) = \prod_{n=1}^{N}\left(\frac{k+\alpha_n^\ell}{k-\alpha_n^\ell}\right)\left(\frac{k-\beta_n^\ell}{k+\beta_n^\ell}\right) \tag{6}$$

Here α_n^ℓ are N simple poles situated both in the upper-half and lower-half k plane, while β_n^ℓ are N simple poles situated only in the lower-half k plane [24, 25]. For uniformity, one may use the same symbols to represent all the roots and singularities of the scattering matrix. In this case the parametrization takes the following form:

$$S_\ell(k) = \prod_{n=1}^{M} \frac{k + \alpha_n^\ell}{k - \alpha_n^\ell}, \quad \text{where} \quad M = 2N \tag{7}$$

This parametrisation ensures conservation of probability current (unitarity of $S_\ell(k)$). It also results in an exact solution to the Fredholm integral equation, as its kernel becomes degenerate; no quadrature is therefore needed [26]. [22] gives an outline of the procedure for transforming the integral equation into a linear system.

3. ΛN elastic scattering

There is very little data on ΛN scattering from experiments. This is because of the difficulty in using free hyperons as projectiles or targets in these experiments, owing to their very short lifetimes (about 10^{-10} s). When compared to free protons that do not decay, the extremely short lifetime of free hyperons poses enormous difficulties. Cross sections for the Λp elastic scattering reaction

$$\Lambda + p \rightarrow \Lambda + p \tag{8}$$

have been reported for laboratory-frame Λ momenta up to a few hundreds of GeV/c. Most of the Λp scattering data obtained have large error bars, and a low number of scattering events, insufficient to constraint the ΛN interaction. The use of such high-momentum Λ beams has significantly increased the Λ decay lengths from a few millimetres to several centimetres, thereby increasing the chances of scattering events in recent experiments. However, the higher the incident momenta, the greater the number of partial waves that must be accounted for; usually, the s-waves only suffice at low incident momenta. In an experimental set up Λn scattering is not as manageable as Λp scattering, due to the non-existence of free stable neutron targets. The lifetime of a free neutron is only approximately 881.5 s.

The link between quantum scattering theory and two of the observables that are usually measured in scattering experiments, differential cross section ($d\sigma/d\Omega$) and integrated cross section (σ), is established by the scattering amplitude, $f_k(\theta)$. The scattering amplitude accounts for the distortion suffered by the incoming wave after scattering. In terms of partial wave analysis, the scattering amplitude may be written in the Faxen-Holtzmark formalism [27] as

$$f_k(\theta) = \frac{1}{2ik} \sum_{\ell=0}^{\infty} (2l+1)(S_\ell(k) - 1) P_\ell(\cos\theta) \tag{9}$$

where P_ℓ is a Legendre polynomial at scattering angle θ. Solving the Radial Schrödinger Equation, one gets the phase shift, which is directly related to the scattering amplitude and cross sections. Using phase shift analysis, the contribution of each partial wave may be independently extracted. A poor phase shift analysis may therefore compromise the inversion procedure. This is an issue which is not of interest in this paper.

For our construction we have used as input theoretical $^1\mathrm{S}_0$ and $^3\mathrm{S}_1$ phase shifts computed by [28] for the NSC97f potential. Even after NN elastic scattering data became widely available, theoretical NN data continued playing an important role in our understanding of the NN force. It is hoped that this strategy will throw more light on the nature of baryon-baryon interactions in the single strangeness sector.

IOP Conf. Series: Journal of Physics: Conf. Series **915** (2017) 012004 doi:10.1088/1742-6596/915/1/012004

4. Results

The results are shown for the s-wave Λp (Figures 1(a) and 1(b)) and Λn (Figures 2(a) and 2(b)) potentials, both in the $S = 0$ and $S = 1$ states. The general features of a baryon-baryon interaction are present in these potentials, except for the $\Lambda n(^3\mathrm{S}_1)$ potential where the short range repulsion is absent. The well-known result, that the $\Lambda N(^3\mathrm{S}_1)$ potential is weaker than the $\Lambda N(^1\mathrm{S}_0)$ potential, is also verified. However, an important feature of these potentials is that the strongest attraction occurs at a smaller radial distance than with most other lambda-nucleon potentials.

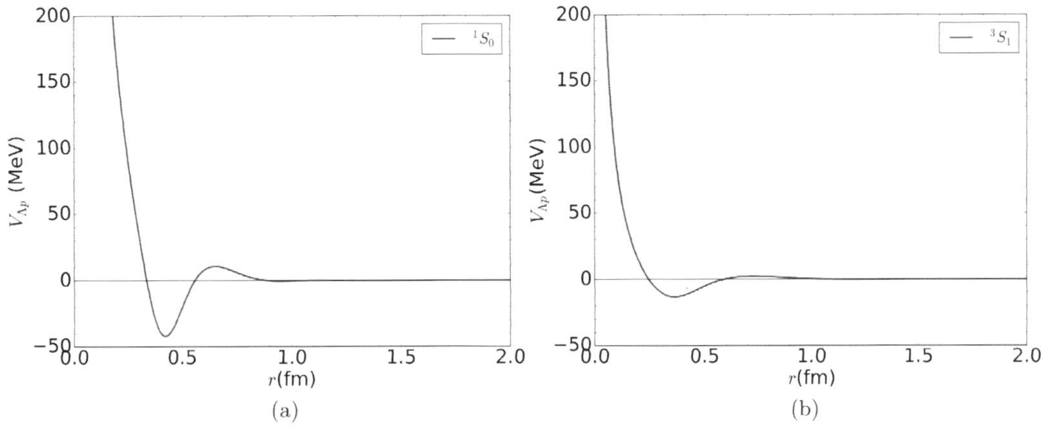

Figure 1: Λp potential in (a) $^1\mathrm{S}_0$ state and (b) $^3\mathrm{S}_1$ state.

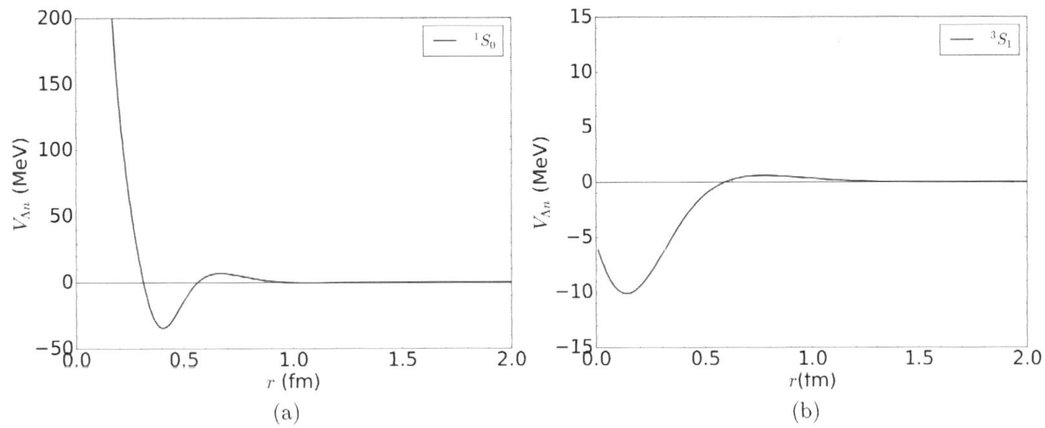

Figure 2: Λn potential in (a) $^1\mathrm{S}_0$ state and (b) $^3\mathrm{S}_1$ state.

Poor choices of $S_\ell(k)$ parametrisation have been known to result in unphysical oscillations in inversion potentials, even in cases where phase shifts from experiments are used [29]. These oscillations are absent in the results obtained here. However, the presence of a small repulsion barrier, which is quite negligible in the spin triplet states, *may* still be a pathology arising from the parametrisation. Since Λ is a neutral elementary particle, this barrier could not possibly be a Coulomb repulsion. Further investigation is needed to clarify the effects of various $S_\ell(k)$ interpolations. This computational artefact is *not* expected to detract from the overall usefulness of our potential. This may be better appreciated in light of the inadequacies of the current potential models used in quantum-mechanical few-body calculations.

IOP Conf. Series: Journal of Physics: Conf. Series **915** (2017) 012004 doi:10.1088/1742-6596/915/1/012004

5. Conclusions

New spin singlet and spin triplet state potentials for the Λp and Λn interactions have been constructed for $\ell = 0$ through Marchenko theory, a quantum inverse scattering formalism. These potentials are energy-independent, a feature that makes them ideal for few-body calculations.

References

[1] Rijken T A, Stoks V G J and Yamamoto Y 1999 *Phys. Rev. C* **59**(1) 21–40
[2] de Swart J J, Nagels M M, Rijken T A and Verhoeven P A 1971 *Springer Tracts in Modern Physics* vol **60** ed Höhler G (Berlin, Heidelberg: Springer Berlin Heidelberg) pp 138–203 *Extended version of the invited paper presented by the first author at the Ruhestein Meeting on 'Low-Energy Hadron Interactions', 1970.*
[3] de Swart J J, Klomp R A M M, Rentmeester M C M and Rijken T A 1996 *Few-Body Problems in Physics '95: In memoriam Professor Paul Urban* ed Guardiola R (Vienna: Springer Vienna) pp 438–447
[4] Hiyama E, Motoba T, Rijken T A and Yamamoto Y 2010 *Progr. Theor. Phys. Suppl.* **185** 1
[5] Nakamoto C, Suzuki Y and Fujiwara Y 1993 *Phys. Lett. B* **318** 587 – 591
[6] Fujiwara Y, Nakamoto C and Suzuki Y 1996 *Phys. Rev. C* **54**(5) 2180–2200
[7] Fujiwara Y, Nakamoto C and Suzuki Y 1996 *Phys. Rev. Lett.* **76**(13) 2242–2245
[8] Chadan K and Sabatier P C 1977 *Inverse problems in quantum scattering theory* (New York: Springer-Verlag)
[9] Kukulin V I and Mackintosh R S 2004 *J. Phys. G Nucl. Part. Phys.* **30**
[10] Fiedeldey H, Sofianos S A, Papastylianos A, Amos K and Allen L J 1990 *Phys. Rev. C* **42**(1) 411–415
[11] Levin B J 1964 *Distribution of zeros of entire functions* (American Mathematical Society) volume Five, Translation of Mathematical Monographs
[12] Marchenko V A 2006 *Mathematical Events of the Twentieth Century* ed Bolibruch A A, Osipov Y S, Sinai Y G, Arnold V I, Bolibruch A A, Vershik A M, Manin Y I, Osipov Y S, Sinai Y G, Tikhomirov V M, Faddeev L D and Philippov V B (Berlin, Heidelberg: Springer Berlin Heidelberg) pp 145–162
[13] Gel'fand I M and Levitan B M 1951 *Izv. Akad. Nauk SSSR Ser. Mat.* **15** 309 – 360
[14] Newton R G and Jost R 1955 *Il Nuovo Cimento* **1** 590 – 622
[15] Agranovich Z S and Marchenko V A 1963 *The inverse problem of scattering theory* (New York: Gordon and Breach Publishers)
[16] Levitan B M 1987 *Inverse Sturm-Liouville Problems* (Utrecht: VNU Science Press)
[17] Deift P and Trubowitz E 1979 *Commun. Pur. Appl. Math.* **32** 121 – 251
[18] Newton R G 1991 *J. Math. Phys.* **32**
[19] Benn J and Scharf G 1969 *Nucl. Phys. A* **134** 481 – 504
[20] Benn J and Scharf G 1972 *Nucl. Phys. A* **183** 319 – 336
[21] Coz M, Kuberczyk J and von Geramb H V 1987 *Z. Phys. A At. Nucl.* **326** 345–351
[22] Kirst T, Amos K, Berge L, Coz M and von Geramb H V 1989 *Phys. Rev. C* **40**(2) 912–923
[23] Sparenberg J M and Baye D 1997 *Phys. Rev. C* **55**(5) 2175–2184
[24] Massen S E, Sofianos S A, Rakityansky S A and Oryu S 1999 *Nucl. Phys. A* **654** 597 – 611
[25] Rakityansky S A, Sofianos S A and Elander N 2007 *J. Phys. A Math. Theor.* **40** 14857
[26] Sofianos S A, Papastylianos A, Fiedeldey H and Alt E O 1990 *Phys. Rev. C* **42**(2) R506–R509
[27] Joachain C J 1983 *Quantum collision theory* 3rd ed (The Netherlands: North-Holland Publishing Company)
[28] Rentmeester M C M and Klomp R A M NN Online http://nn-online.org/ accessed: 2017 May 01
[29] Howell L L, Sofianos S A, Fiedeldey H and Pantis G 1993 *Nucl. Phys. A* **556** 29 – 41

IOP Conf. Series: Journal of Physics: Conf. Series **915** (2017) 012005 doi:10.1088/1742-6596/915/1/012005

Few-body integrodifferential equation on Lagrange mesh

G J Rampho, L C Mabunda and M Ramantswana

Department of Physics, University of South Africa, P.O. Box 392, Pretoria 0003, South Africa

E-mail: ramphjg@unisa.ac.za

Abstract. The two-variable integrodifferential equation for few-body systems is solved using the Lagrange-mesh method. The method transforms the equation into a system of algebraic equations that are solved as a non-symmetric matrix eigenvalue problem. Convergence properties of the solution to the integrodifferential equation in relation to the problem parameters is investigated. The accuracy of the converged solution is tested by calculating the binding energies and root-mean-square radii of selected few-body systems. The results are compared to those generated by other methods.

1. Introduction

The integrodifferential equation (IDE) approach is one of the several techniques devised to simplify the many-body Schrödinger equation for ease of numerical solution. The approach is based on the Faddeev formalism where the many-body Schrödinger wave function is decomposed into a sum of two-body amplitudes that depend on two global variables of the system [1]. The formalism also decomposes the Schrödinger equation into a set of coupled IDEs, for the amplitudes, that can be solved numerically with comparatively less complexity than the original Schrödinger equation. Such a decomposition explicitly accounts for two-body correlations in the system. All the IDEs are identical for systems of identical particles. Therefore, for such systems, only one IDE needs to be solved. The numerical solutions to the equations have been shown to be as accurate as other approaches to the solution of the many-body Schrödinger equation. However, unlike in the other approaches, the form of the IDEs in this approach does not depend on the size of the system.

The numerical solution of the IDEs has proved challenging because of their characteristic oscillatory non-local projection kernels. So far the two-variable integrodiferential equations have been solved using adiabatic approximation techniques [1, 2, 3, 4] that decouples the equation into two single-variable equations. The challenge with the adiabatic approximation is that the energies of the systems are often extrapolated from the calculated ones [5]. This raises concerns, that have not yet been addressed, about the accuracy of the calculated amplitudes. Another, not so popular solution method for the IDEs, is the perturbation method [6, 7], that solves the equations iteratively. In this method, the zeroth order (unperturbed) differential equations are separable in the two variables and can both be readily solved. Both the adiabatic approximation and the perturbation solutions of the IDEs become generally less reliable for few-body systems, and particularly for systems involving hard-core interactions [6]. In this paper the Lagrange-mesh method [8] is explored for the direct solution of the few-body IDEs, for the first time.

IOP Conf. Series: Journal of Physics: Conf. Series **915** (2017) 012005 doi:10.1088/1742-6596/915/1/012005

The Lagrange-mesh method is a variational method that employs basis functions that are constructed on a discretized problem domain. The basis functions so constructed have specialized properties, like orthonormality at the chosen discrete points, which leads to banded matrices for the Hamiltonian of the problem. This method was shown to generate converged solutions for the integral Schrödinger equation in momentum space [9], which involve non-local kernels similar to the few-body IDEs. The Lagrange-mesh method is widely discussed in the literature [8, 9, 10, 11, 12, 13, 14] and is here explored for solving the few-body IDEs.

This paper is organized as follows. In section 2 the Faddeev formalism is explained and the IDEs for few-body systems are outlined. Section 3 describes the construction of the Lagrange mesh matrix elements for the IDEs. Convergence tests of the numerical solutions are discussed in Section 4 and conclusions are reported in Section 5.

2. The few-body integrodifferential equation

Consider a system of A particles of equal mass m and position vector $\vec{r}_i$, $i = 1, \cdots, A$, in which two-body interactions $V(\vec{r}_{ij})$, where $\vec{r}_{ij} = \vec{r}_i - \vec{r}_j$, are dominant. The Schröndiger equation for the many-body system has the form

$$\left[-\sum_{i}^{A} \frac{\hbar^2}{2m_i}\nabla_i^2 + \sum_{ij}^{A} V(\vec{r}_{ij}) - E\right] \Psi(\vec{x}) = 0 \tag{1}$$

where $\vec{x} \equiv (\vec{r}_1, \vec{r}_2, \cdots, \vec{r}_A)$, E the energy and $\Psi(\vec{x})$ the total wave function of the system. In the Faddeev formalism, the solution to this many-body problem is constructed by decomposing the many-body Schrödinger wave function in the form [15]

$$\Psi(\vec{x}) = \sum_{ij}^{A} \psi_{ij}(\vec{x}) \tag{2}$$

where $\psi_{ij}(\vec{x})$ are Faddeev two-body amplitudes. These amplitudes depend on all the degrees-of-freedom of all the particles in the system and satisfy $\frac{1}{2}\,A\,(A-1)$ coupled equations of the form

$$[\hat{T} - E]\psi_{ij}(\vec{x}) = -V(\vec{r}_{ij})\,\Psi(\vec{x})\,. \tag{3}$$

where $\hat{T}$ represents the kinetic energy operator in Eq. (1). All the equations for (3) the amplitudes $\psi_{ij}(\vec{x})$ add up to the complete Schrödinger equation for the system. The realization of numerical solutions to (3) requires the introduction of some approximations to the equation.

The system is characterized by a number of global variables, two of which are the hyperradius $r \in [0, \infty]$ determined by the magnitude of all the relative position vectors of the particles in, and $z \in [-1, +1]$ related to one of the hyperangles of the system. These two variables are related by $r_{ij} = r\sqrt{\frac{1}{2}(1+z)}$. The $\frac{1}{2}\,A(A-1)$ coupled IDEs for the system (3), in these two global variables, are all identical and can be written in the form [16]

$$\begin{aligned} -\frac{\hbar^2}{m}\left[\frac{\partial^2}{\partial r^2} + \frac{1}{r^2}\left(4\,\hat{T}_z - \frac{2\,\ell(\ell+1)}{1-z} - \frac{(3A-4)(3A-6)}{4}\right)\right] F_\ell(r,z) \\ + \Big[V(r,z) - E\Big]\,F_\ell(r,z) = -V(r,z)\int_{-1}^{+1} f_\ell(z,z')\,F_\ell(r,z')\,\mathrm{d}z' \end{aligned} \tag{4}$$

where $F_\ell(r,z)$ are the reduced Faddeev two-body amplitudes and ℓ is the relative orbital angular momenta of the interacting pair. In Eq. (4), the kinetic energy for the center-of-mass of the system is removed, and the abbreviation

$$\hat{T}_z = \frac{1}{w_0(z)} \frac{\partial}{\partial z} (1-z)^2 \, w_0(z) \frac{\partial}{\partial z} \tag{5}$$

is used. The weight function $w_0(z) = (1-z)^{\alpha_0}(1+z)^{\beta_0}$ is defined by the parameters $\alpha_0 = (3A-8)/2$ and $\beta_0 = \ell + 1/2$, which are specified by the problem of interest. The kernel $f_\ell(z, z')$ projects the angular variables for the other pairs of particles into that of the focus pair. This kernel is defined in terms of the Jacobi polynomials $P_K^{\alpha,\beta}(z)$ with normalization $h_K^{\alpha,\beta}$ by [15]

$$f_\ell(z,z') = w^J(z') \sum_{K=0}^{\infty} \frac{q_{K\ell}^{\alpha,\beta}}{h_K^{\alpha,\beta}} P_K^{\alpha,\beta}(z) \, P_K^{\alpha,\beta}(z') \tag{6}$$

where $(\alpha, \beta) = (\alpha_0, \beta_0)$ and $w^J(z) = (1-z)^\alpha (1+z)^\beta$. The coefficients

$$q_{K\ell}^{\alpha,\beta} = \frac{A-2}{P_K^{\alpha,\beta}(1)} \left[(-1)^\ell \, 2^{1-\ell} \, P_K^{\alpha,\beta}(-\tfrac{1}{2}) + \tfrac{1}{2}(A-3) \, P_K^{\alpha,\beta}(-1) \right] \tag{7}$$

are related to pairs of particles that are connected and those that are not connected to the $\{ij\}$ pair. For systems of particles with differing masses, the first term in the square brackets is slightly modified.

Note that (4) represent a set of coupled equations for the amplitudes with different values of ℓ. This set of equations is reduced to one IDE by setting $\ell = 0$. The resulting equation is then modified by adding the zeroth ($K = 0$) and subtracting the first ($K = 1$) order hypercentral potential multipoles,

$$V_K(r) = \frac{1}{h_K^{\alpha,\beta}} \int_{-1}^{+1} V(r,z) \, P_K^{\alpha,\beta}(z) \, w(z) \, \mathrm{d}z \tag{8}$$

from the equation. The zeroth order multipole accounts for the addition of the effects of higher orbital angular momentum [16], while the first order multipole accounts for the removal of the effects of the spurious amplitudes [5]. Only the effects of spurious amplitudes are considered in the present work. As a result, the two-variable $s-$state intgrodifferential equation solved is [15]

$$\begin{aligned} \frac{\hbar^2}{m} \left[\frac{\partial^2}{\partial r^2} + \frac{1}{r^2} \left(4\hat{T}_z - \frac{(3A-4)(3A-6)}{4} \right) \right] F(r,z) \\ + \left[\bar{V}(r,z) - E \right] F(r,z) = -\bar{V}(r,z) \int_{-1}^{+1} f(z,z') \, F(r,z') \, \mathrm{d}z' \end{aligned} \tag{9}$$

where

$$\bar{V}(r,z) = V(r,z) - V_1(r) \, P_1^{\alpha,\beta}(z) \tag{10}$$

and the subscripts on the amplitude are now omitted since they are redundant. In anticipation of later discussions, equation (9) is cast in the eigenproblem form

$$\left[\hat{H}_0 + \bar{V}(r,z) \right] F(r,z) + \bar{V}(r,z) \int_{-1}^{+1} f(z,z') \, F(r,z') \, \mathrm{d}z' = E \, F(r,z) \tag{11}$$

where

$$\hat{H}_0 = -\frac{\hbar^2}{m} \frac{\partial^2}{\partial r^2} - \frac{4\hbar^2}{m\, r^2} \left(\hat{T}_z - \frac{(3A-4)(3A-6)}{16} \right) . \tag{12}$$

For bound-state solutions to (11), which are the focus of this paper, specific boundary conditions need to be satisfied by the amplitudes $F(r,z)$. Such conditions are determined by principles of

IOP Conf. Series: Journal of Physics: Conf. Series **915** (2017) 012005 doi:10.1088/1742-6596/915/1/012005

quantum mechanics involving the requirement for a realistic description of the physical properties of the system.

The form of the interaction potentials in the equation determines the type of boundary conditions to be satisfied by the amplitudes. The asymptotic short- and long-range radial forms of the amplitudes, for bound states, are

$$F(r,z) \stackrel{r\to 0}{\longrightarrow} r^{(3A-4)/2} \quad \text{and} \quad F(r,z) \stackrel{r\to\infty}{\longrightarrow} \mathrm{e}^{-\sqrt{E_n}\,r} \tag{13}$$

where E_n is the energy of one of the states of the system. These asymptotic forms translate into the boundary conditions

$$F(0,z) = F(\infty,z) = 0 \tag{14}$$

which are consistent with suppressing the singularity of (12). The amplitudes are also required to vanish at all other singularities of the Hamiltonian. Since the angular-related part of the kinetic energy operator does not appear to have singularities, the boundary conditions

$$F(r,\pm 1) \neq 0\,, \tag{15}$$

are adopted. That is, if the potential does not have singularities in the z domain, then $F(r,\pm 1)$ are just functions of r. In the next section the numerical solution of (11) is constructed using the Lagrange-mesh method.

3. Lagrange mesh matrix elements

The Lagrange-mesh method is widely discussed in the literature and the reader is referred to [8] for details on this method. In this method, the problem domain $[a,b]$ is segmented into intervals defined by a by a quadrature

$$\int_a^b g(x)\,\mathrm{d}x = \sum_{i=1}^{N} \lambda_i\, g(x_i) \tag{16}$$

where $g(x)$ is a real function, while x_i and λ_i are the quadrature abscissas and weights, respectively. The method is characterized by basis functions $f_j(x)$ that have the properties [8]

$$f_j(x_i) = \lambda_i^{-1/2}\,\delta_{ij} \tag{17}$$

$$\int_a^b f_i(x)\,V(x)\,f_j(x)\,\mathrm{d}x = V(x_i)\,\delta_{ij} \tag{18}$$

$$\int_a^b f_i(x)\,f_j^{(k)}(x)\,\mathrm{d}x = \lambda_i^{1/2}\,f_j^{(k)}(x_i) \tag{19}$$

where $V(x)$ is a real function and $f^{(k)} \equiv d^k f/dx^k$. These properties reduce dynamical equations for quantum mechanical problems to a simple set of algebraic equations that are readily solved.

A simple example application of the Lagrange functions is provided by the one-dimensional Schrödinger equation

$$\left[-\frac{\mathrm{d}^2}{\mathrm{d}x^2} + V(x)\right]\psi(x) = E\,\psi(x)\,; \qquad x \in [a,b] \tag{20}$$

where $V(x)$ is a scalar potential, $\psi(x)$ the wave function, and E the energy of a given system. The Lagrange-mesh solution to this equation is constructed by approximating the wave function as

$$\psi(x) = \sum_{j=1}^{N} c_j\, f_j(x) \tag{21}$$

where c_j are variational parameters and $f_j(x)$ the N Lagrange functions defined on a grid $a \leq x_j \leq b$ (j=1,2,3,...,N). Taking the matrix elements of (20) with the expansion (21) leads to

$$-\left\langle \psi(x) \,|\, \psi''(x) \right\rangle + \left\langle \psi(x) \,|\, V(x) \,|\, \psi(x) \right\rangle == E \, \left\langle \psi(x) \,|\, \right\rangle \,. \tag{22}$$

This equation simplifies, thanks to the properties of the Lagrange functions, to the set of algebraic equations [8]

$$\sum_{j=1}^{N} \left[-\lambda^{1/2} \, f_j''(x_i) + V(x_i) \, \delta_{ij} \right] \, c_j = E \, c_i \,, \tag{23}$$

for the variational parameters c_i. Note that the matrix elements in (23) require only the evaluation of the known functions $f_j''(x)$ and $V(x)$ at the known points x_i. The Lagrange-mesh method is similarly applied to (11), however, this equation presents non-Hermitian and non-local operators, which lead to a non-symmetric matrix eigenvalue problem.

The amplitudes in (11) are approximated with the series

$$F(r,z) = \sum_{i=1}^{K} \sum_{j=1}^{N} C_{ij} \, R_i(r) \, U_j(z) \tag{24}$$

where C_{ij} are variational parameters, $R_i(r)$ and $U_j(z)$ bases functions that satisfy the boundary conditions (14) and (15). These functions are defined on the grid $r_i \, (i = 1, 2, \ldots, K)$ and $z_j \, (j = 1, 2, \ldots, N)$ which are chosen as the roots of the Laguerre and Jacobi polynomials, respectively, of orders K and N. Lagrange bases defined on this grid are the Lagrange-Laguerre and Lagrange-Jacobi functions. The Lagrange-Laguerre functions, regularized by r, employed here have the form [8]

$$R_i(r) = (-1)^i \left[\frac{r_i \, r^2 \, w^L(r)}{h_K^\sigma} \right]^{1/2} \frac{L_K^\sigma(r)}{r - r_i} \tag{25}$$

where h_K^σ is the normalization coefficient of and $w^L(r) = r^\sigma \mathrm{e}^{-r}$ the weight function for the Laguerre polynomials $L_K^\sigma(r)$ of order K. The parameter σ is usually chosen so that (25) displays the required boundary behaviour of the exact wave function at small r. However, here $\sigma = 0$. The regularized Lagrange-Jacobi functions have the form [17]

$$U_j(z) = (-1)^{N-j} \left[\frac{(1-z^2)^\mu \, w^J(z)}{h_N^{\alpha,\beta} (2\,N + \gamma) \, (1 - z_j^2)^{\mu - 1}} \right]^{1/2} \frac{P_N^{\alpha,\beta}(z)}{z - z_j} \tag{26}$$

where μ is the regularizing parameter and $\gamma = \alpha + \beta + 1$. The values of the parameters (α, β) are not necessarily the same as those specified in the previous section. However, when this is the case, then the corresponding polynomials $P_N^{\alpha,\beta}(z)$ constitute the eigenfunctions of the angular component of the kinetic energy operator. In general, these parameters need to be optimized for a given potential.

Using the expansion (24) in (11) with (25) and (26), and constructing the matrix elements for the equation, leads to the matrix eigenvalue problem

$$\sum_{kl} \left\{ \frac{1}{h^2} \, H_{ij,kl}^0 + \bar{V}(h \, r_i, z_l) \, \left[\delta_{jl} + \sqrt{\lambda_j \, \lambda_l} \, f(z_l, z_j) \right] \, \delta_{ik} \right\} \, C_{kl} = E \, C_{ij} \tag{27}$$

for the variational parameters, where $\lambda_i = w_i^J / w^J(z_i)$ (w_i^J are the quadrature weights), h a scaling parameter and

$$H_{ij,kl}^0 = \frac{\hbar^2}{m} \left[T_{ik}^r \, \delta_{jl} - \frac{4}{r_i^2} \left(T_{jl}^z - \frac{(3A-4)(3A-6)}{16} \, \delta_{jl} \right) \delta_{ik} \right] . \tag{28}$$

The matrix elements T^r_{ij} are those given in [8] while T^z_{ij} are taken from [18]. It should be noted that the matrix elements in the eigenvalue problem (27) are functions of the quadrature weights and abscissas, as well as the free parameters h, α, β and μ. Convergence properties of the solutions to (27) with variations in the bases sizes are discussed in the next section.

4. Convergence tests

The evaluation of the matrix elements discussed in the previous section require the specification of the parameters $\{h, \mu, \alpha, \beta\}$, as well as the roots of the Laguerre and Jacobi polynomials. The mesh was based on Gaussian quadrature, which were determined using standard routines [19] with a tolerance of 3×10^{-14}. The regularisation parameter μ is determined by the boundary conditions (15). The value $\mu = -\frac{1}{2}$ was adopted. The scaling parameter h is determined by the hyperradial range of the total potential of the system [20]. Each particle in the system experiences an average hyperradial potential described by the hypercentral potential $V_0(r)$. Figure 1 show the three-body to six-body hypercentral potentials described by (8). As can be seen in the figure, the range of this hypercentral potential depends on the size of the system. In this work, the scaling parameter was chosen such that $h = r_m/x_N$ where r_m is the range of the hypercentral potential and x_N the last zero of the Laguerre polynomial. The range $r_m = 20\,\text{fm}$ was used for the three-body and four-body, while $r_m = 30\,\text{fm}$ was used for the five-body and six-body systems. The parameters of the functions $U_j(z)$ were set to $(\alpha, \beta) = (\alpha_0, \beta_0)$, which are related to the eigenfunctions of the angular kinetic energy operator.

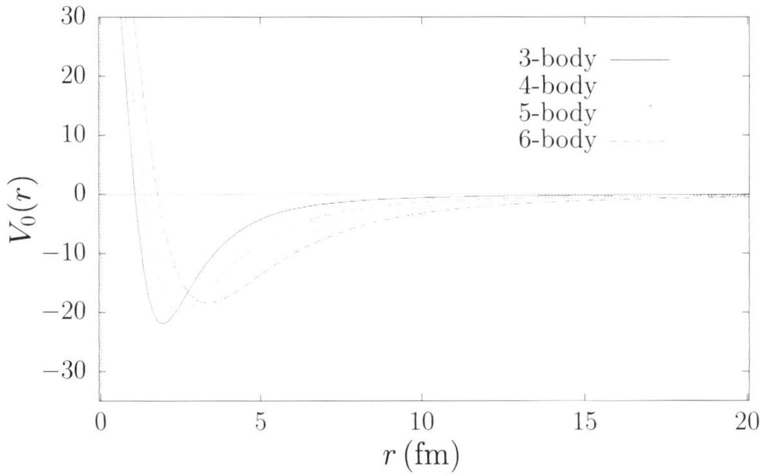

Figure 1. (Color online) Variation of the range of few-body hypercentral potentials with the size of the system.

Test applications of the matrix elements were conducted with the central Volkov potential [21] and the Ali-Bodmer potential [22] for three-, four-, five- and six-body systems. The variation of the ground-state energies of the systems with the bases sizes were investigated. The investigations were restricted only to the cases of equal radial and angular bases size $N = K$. In addition, a restriction of $M \geq N$, on the number of terms M in the series in (6), was imposed, based on tests results of the projection function on polynomial and exponential functions.

4.1. Three-body systems

In the first case, the three-body s-state IDE was solved for the $3n$ ($\hbar^2/m = 41.47\,\text{MeV} \cdot \text{fm}^{-2}$) and 3α ($\hbar^2/m = 10.366\,\text{MeV} \cdot \text{fm}^{-2}$) systems. The convergence of the ground state energies was compared with that of the Faddeev equations for the systems, solved in the same way. The Faddeev E_{Fad} and IDE E_{IDE} results are given in table 1. As can be seen from the table, the Faddeev results converge rapidly to -8.431 MeV for the $3n$ system and to $-4.141\,MeV$ for the 3α system. These results coincide with those reported in the literature [23, 24]. The IDE results, on the other hand, converge very slowly to -8.111 MeV and -3.302 MeV for the Volkov and Ali-Bodmer potentials, respectively. These results are about 5% and 20% less than those of the Faddeev for the two potentials. There are several possible causes of these discrepancies, one of which is the effects of the spurious potential [25].

Table 1. The calculated ground-state energies E_0 (MeV) for the three-nucleon ($3\,n$) system and the $3\,\alpha$ system as a function of the basis size $N^2 = N_r \otimes N_z$.

	$3\,n$		$3\,\alpha$	
N	E_{Fad}	E_{IDE}	E_{Fad}	E_{IDE}
10	-8.410 23	-8.496 29	-7.469 91	-7.571 26
20	-8.430 91	-8.177 31	-4.140 74	-3.302 57
30	-8.430 93	-8.137 88	-4.140 71	-3.301 92
40	-8.430 93	-8.123 90	-4.140 71	-3.301 74
50	-8.430 93	-8.117 36	-4.140 71	-3.301 65
60	-8.430 93	-8.113 78	-4.140 71	-3.301 60
70	-8.430 93	-8.111 61	-4.140 71	-3.301 57
	-8.430 93[a]		-4.140 71[b]	

[a] Reference [23], [b] Reference [24]

4.2. *Many-body systems*

To investigate the effects of the spurious potential on the results, ground state energies for A-body systems, where $A = 3, 4, 5, 6$, were calculated with the spurious potential eliminated. The E_{IDE} results are shown in table 2 and table 3 for the Volkov and Ali-Bodmer potential, respectively. The energies for all the systems converge quite slowly, and, as indicated in [25], do not appear to be significantly affected by the spurious potential. There is a general discrepancy of about 11% in the Volkov results compared to the s-wave results of the hyperspherical harmonics method [23] for the same potential. However, the hyperspherical harmonics results are obtained with bases sizes in the orders of millions, for six-body system, for example.

The quality of the wave functions is often tested by calculating root-mean-square radii of the systems studied. The results for the root-mean-square radii of three-body and four-body systems are presented in table 4 for the Volkov and Ali-Bodmer potentials. The results for the 3α system are about 12% higher than those reported in the literature [24].

5. Conclusion

The Lagrange-mesh method was explored for solving the few-body IDEs. Matrix elements for the non-Hermitian and non-local operators of the equations were constructed using regularized Lagrange-Laguerre and Lagrange-Jacobi functions. The resulting non-symmetric matrix was first solved with standard nuclear potentials for three-body systems. To investigate the effect of the non-symmetric form of the eigenproblem on the convergence of the numerical solution,

Table 2. Calculated ground-state energies E_0 for A-body systems with the Volkov potential as a function of the basis size $N^2 = N_r \otimes N_z$.

	E_0 (MeV)			
N	$A=3$	$A=4$	$A=5$	$A=6$
10	-8.484 43	-37.774 91	-80.186 69	-134.889 84
20	-8.162 19	-29.802 67	-82.847 96	-146.360 81
30	-8.132 75	-27.487 53	-69.728 30	-125.455 83
40	-8.121 71	-26.858 43	-64.512 53	-116.858 91
50	-8.116 26	-26.629 04	-62.437 31	-113.429 32
60	-8.113 16	-26.526 83	-61.490 23	-111.863 80
70	-8.111 22	-26.474 50	-61.004 19	-111.060 43
[23]	-8.430 9	-30.252	-68.280	-122.776

Table 3. Same as in Table 2 for the Ali-Bodmer potential.

	E_0 (MeV)			
N	$A=3$	$A=4$	$A=5$	$A=6$
10	-7.609 33	-8.191 16	-9.116 88	-15.654 54
20	-3.335 48	-5.694 18	-7.817 98	-10.331 76
30	-3.310 00	-5.739 99	-8.121 78	-10.615 47
40	-3.304 83	-5.748 70	-8.167 24	-10.649 52
50	-3.303 09	-5.751 76	-8.180 21	-10.659 22
60	-3.302 26	-5.753 33	-8.186 09	-10.664 49
70	-3.301 78	-5.753 99	-8.189 62	-10.666 71

Table 4. Root-mean-square radii for A-body systems with $A = 3, 4, 5, 6$ for the Volkov (V) and Ali-Bodmer (AB) potentials, calculated with $N = 70$.

A	$\langle r^2 \rangle_{\rm V}^{1/2}$	$\langle r^2 \rangle_{\rm AB}^{1/2}$
3	1.931	2.725
4	1.238	2.335

the s-wave Faddeev equations and the s-state IDE were solved. The results showed rapid convergence for the solutions, with the basis size, of the Faddeev equations and a very slow convergence for the corresponding IDE. Moreover, the Faddeev solutions converged to results reported in the literature, obtained using other solution methods. However, the IDE solutions converged to results that are less than the Faddeev results, and different from those obtained using other solution methods, by about 11%. Based on the results for the Faddeev solutions, the non-symmetric form for the IDE is not responsible for the slow convergence of the solution. Therefore, the convergence problems could emanate from the form of the basis functions and/or the implementation of boundary conditions of the problem.

The IDE was also solved for the ground-state energies of A-body systems where $A = 3, 4, 5, 6$, with the contribution of spurious potential removed. Ground state energies and root-mean-square radii of the $A = 3, 4$ systems were calculated. The results showed very slow convergence with the basis sizes, to values $\sim 11\%$ less than those obtained using the hyperspherical harmonics method. Both the ground-state energies and root-mean-square radii of the systems were poorly determined, reflecting the possible poor quality of the numerical wave functions. The convergence of the IDE Lagrange-mesh solutions, albeit slow, and the rapid convergence of the Faddeev Lagrange-mesh solutions, indicate that the method is a promising tool to generate direct solutions to the IDE. Further investigations on the effectiveness of the Lagrange-mesh solution of the IDE are necessary. Work is underway to improve the numerical treatment of the few-body IDE.

References

[1] de la Ripelle M F, Fiedeldey H and Sofianos S A 1988 *Phys. Rev. C* **38** 449 – 466
[2] Das T K and Roy S 1991 *Pramana- Journal of Physics* **36** 305 – 312
[3] de la Ripelle M F 2010 *Nucl. Phys. A* **839** 11 – 41
[4] de la Ripelle M F 2016 *Phys. Lett. B* **753** 1 – 7
[5] de la Ripelle M F, Fiedeldey H and Sofianos S A 1989 *Few-Body Syst.* **6** 157 – 174
[6] Das T K and Chattopadhyay R 1993 *Fizika B* **2** 73 – 81
[7] Barnea N and Viviani M 2000 *Phys. Rev. C* **61**(3) 034003
[8] Baye D 2015 *Phys. Rep.* **565** 1 – 107
[9] Lacroix G, Semay C and Buisseret F 2012 *Phys. Rev. E* **86**(2) 026705
[10] Baye D, Hesse M and Vincke M 2002 *Phys. Rev. E* **65**(2) 026701
[11] Baye D 2006 *physica status solidi (b)* **243** 1095–1109
[12] Vincke M, Malegat L and Baye D 1993 *J. Phys. B: Atom. Mol. Opt. Phys.* **26**
[13] Baye D and Vincke M 1999 *Phys. Rev. E* **59**(6) 7195–7199
[14] Baye D 1995 *J. Phys. B: Atom. Mol. Opt. Phys.* **28** 4399
[15] Fabre de la Ripelle M 1986 *Few-Body Syst.* **1** 181–192
[16] de la Ripelle M F 1983 *Ann. Phys.* **147** 281 – 320
[17] Rampho G J 2016 *J. Phys.: CS* **49** 27501
[18] Rampho G J 2016 *J. Phys. A: Math. Theor.* **49** 27501
[19] Press W H, Teukolsky S A, Vetterling W T and Flannery B P 1992 *Numerical Recipes in FORTRAN (2nd Ed.): The Art of Scientific Computing* (New York, USA: Cambridge University Press) ISBN 0-521-43064-X
[20] Masui H, Aoyama S and Baye D 2013 *Prog. Theor. Exp. Phys.* **2013** 123A02
[21] Volkov A 1965 *Nucl. Phys.* **74** 33 – 58
[22] Ali S and Bodmer A 1966 *Nucl. Phys.* **80** 99 – 112
[23] Gattobigio M, Kievsky A and Viviani M 2011 *Phys. Rev. C* **83**(2) 024001
[24] Papp Z and Plessas W 1996 *Phys. Rev. C* **54**(1) 50 56
[25] de la Ripelle M F and Larsen S Y 1992 *Few-Body Syst.* **13** 199 – 206

IOP Conf. Series: Journal of Physics: Conf. Series **915** (2017) 012006 doi:10.1088/1742-6596/915/1/012006

Artificial neural networks with an infinite number of nodes

K Blekas and I E Lagaris

Department of Computer Science and Engineering University of Ioannina, Ioannina 45500, Greece

E-mail: kblekas@cs.uoi.gr, lagaris@cs.uoi.gr

Abstract. A new class of Artificial Neural Networks is described incorporating a node density function and functional weights. This network containing an infinite number of nodes, excels in generalizing and possesses a superior extrapolation capability.

1. Introduction

Artificial Neural Networks (ANNs) are known to be universal approximators [1, 2] and naturally have been applied to both regression and classification problems. Usual types are the so called feed-forward "Multi-Layered Perceptrons (MLP)" and the "Radial Basis Function (RBF) networks". The usefulness of ANNs is indisputable and there is a vast literature describing applications in different fields, such as pattern recognition and data fitting [3, 4], the solution of ordinary and partial differential equations [5, 6, 7], stock price prediction [8], etc. ANNs can learn from existing data the underlying law (i.e. a function) that produced the dataset. The network parameters, commonly referred to as "weights", are adjusted by a "training" procedure so as to represent as closely as possible the generating function. When the data are outcomes of experimental measurements, they are bound to be corrupted by systematic errors and random noise, a fact that further complicates the "learning" task.

"Over-training" is a phenomenon where, although a network may fit a given dataset accurately, it fails miserably when used for interpolation. To evaluate the performance of a network, the dataset is usually split into the "Training-set" and the "Test-set". The ANN is trained using the former, and it is evaluated by examining its interpolation capability on the latter. If the interpolation is satisfactory, the network is said to "generalize" well, and hence it may be trusted for further use. Over-trained networks do not generalize well. An empirical observation is that given two networks with similar performance on the training set, the one with fewer parameters is likely to generalize better. Several techniques have been developed aiming to improve the network's generalization; for instance "weight decay" and "weight bounding" [9, 10], "network pruning" [11], etc. These methods make use of an additional set, the validation set, i.e. they partition the data into three subsets; the training, the validation and the test subset. Training may yield in several candidate models (i.e. networks with different weight values). Then the validation set is used to pick the best performing model, and finally the test set will give an estimate for the size of the expected error on new data.

Another framework for modeling is through sparse supervised learning models [4, 12] which utilize only a subset of the training data, discarding unnecessary samples based on certain criteria. Sparse models widely used are, among others, the Lasso [13], the Support Vector Machines (SVMs) [14] and the Relevance Vector Machines (RVMs) [15]. Sparse learning applies either L_1 regularization, leading to penalized regression schemes [13], or sparse priors on the model parameters under the Bayesian framework [15, 16, 17, 18]. Quite recently, Toronto's "Dropout" technique [19] has appeared, stressing the importance of the subject and indicating the current keen interest of the scientific community. This approach employs a stochastic procedure for node removal during the training phase, thus reducing the number of model parameters, in an effort to avoid over-training.

In the present article we introduce a new type of neural network the FWNN, with weights that depend on a continuous variable, instead of the traditional discrete index. This network employs a small set of adjustable parameters and at the same time presents superior generalization performance as indicated by an ample set of numerical experiments. In addition to the interpolation capability we studied the extrapolation properties of FWNN, found to be quite promising. Presently, and in order to illustrate the new concept, we focus on particular forms of the involved weight and kernel functions, which however are not restrictive by any means. The idea of replacing the indexed parameters with continuous functions has been previously considered in [20], in a different setting and with rather limited implementation prospects. However since then, no further followups have been spotted in the literature.

In section 2, we introduce the proposed neural network with continuous weight functions, by associating it to an ordinary radial basis function network and presenting the process of the transition to the continuum. In section 3, we report comparative outcomes of numerical experiments conducted on both homemade datasets and on well known benchmarks established in the literature. Finally, in section 4, we summarize the strengths of the method, and consider different architectural choices that may become the subject of future research.

2. A new type of neural network

Radial basis functions (RBF) are known to be suitable for function approximation and multivariate interpolation [21, 22]. An RBF network with K Gaussian activation nodes, may be written as:

$$N_{RBF}(\mathbf{x};\theta) = w_0 + \sum_{j=1}^{K} w_j \phi(\mathbf{x};\mu_j,\sigma_j) = w_0 + \sum_{j=1}^{K} w_j \exp\left(-\frac{|\mathbf{x}-\mu_j|^2}{2\sigma_j^2}\right) \tag{1}$$

where $\mathbf{x}, \mu_j \in R^n$ and $\theta = \{w_j, \mu_j, \sigma_j\}_{j=0}^{K}$ collectively denote the network parameters to be determined via the training procedure. The total number of adjustable parameters is given by the expression

$$N_{var}^{RB} = K(2+n) + 1 \tag{2}$$

which grows linearly with the number of network nodes.

Consider a dataset $S = \{\mathbf{x}^{(l)}, f^{(l)}\}$, where $f^{(l)}$ is the desired output for the corresponding input $\mathbf{x}^{(l)}$, and let $T \subset S$ be a subset of S. The interpolating RBF network is then determined by minimizing the mean squared deviation over T:

$$E_{[T]}(\theta) \stackrel{\text{def}}{=} \frac{1}{\#T} \sum_{\mathbf{x}^{(l)}, f^{(l)} \in T} \left(N_{RBF}(\mathbf{x}^{(l)};\theta) - f^{(l)}\right)^2 \tag{3}$$

Let $\widehat{\theta} = \{\widehat{w}_j, \widehat{\mu}_j, \widehat{\sigma}_j\}_{j=1}^{K}$ be the minimizer of $E_{[T]}(\theta)$, i.e.

$$\widehat{\theta} = \arg\min_{\theta}\{E_{[T]}(\theta)\} \ . \tag{4}$$

The network's generalization performance is measured by the mean squared deviation, $E_{[S-T]}(\widehat{\theta})$, over the relative complement set $S-T$. In the neural network literature, T is usually referred to as the *"training"* set, while $S-T$ as the *"test"* set. A well studied issue is the proper choice for K, the number of nodes.

The training "error" $E_{[T]}(\widehat{\theta})$, is a monotonically decreasing function of K, while the test "error" $E_{[S-T]}(\widehat{\theta})$, is not. Hence we may encounter a situation where adding nodes, in an effort to reduce the training error, will result in increasing the test error, spoiling therefore the network's generalization ability. This is what is called "over-fitting" or "over-training", which is clearly undesirable. An analysis of this phenomenon, coined under the name *"bias-variance dillema"* may be found in [23]. Over-fitting is a serious problem and considerable research effort has been invested to find ways to deter it, leading to the development of several techniques such as model selection, cross-validation, early stopping, weight decay, and weight pruning [4, 12, 23, 24].

2.1. Functionally Weighted Neural Network (FWNN)

We propose a new type of neural network by considering an infinite number of hidden nodes. This is facilitated by introducing a node density function $\rho(s) = \frac{1}{1-s^2}$ of a continuous variable $s \in [-1, 1]$ with diverging integral $\int_{-1}^{1} \rho(s)ds \to \infty$. We define the "Functionally Weighted Neural Network" in correspondence to (1) as:

$$N_{FW}(\mathbf{x};\theta) = \int_{-1}^{1} \frac{ds}{1-s^2} \tilde{w}(s) \exp\left(-\frac{|\mathbf{x}-\mu(s)|^2}{2\sigma^2(s)}\right) , \tag{5}$$

by applying the following transitions:

$$w_j \longrightarrow \tilde{w}(s),\ \mu_j \longrightarrow \mu(s),\ \sigma_j \longrightarrow \sigma(s),\ \sum_{j=1}^{K} \longrightarrow \int_{-1}^{1} \frac{ds}{1-s^2} \tag{6}$$

In order to use the Gauss–Chebyshev quadrature, we reset $\tilde{w}(s) = \sqrt{1-s^2}w(s)$ and the network becomes:

$$N_{FW}(\mathbf{x};\theta) = \int_{-1}^{1} \frac{ds}{\sqrt{1-s^2}} w(s) \exp\left(-\frac{|\mathbf{x}-\mu(s)|^2}{2\sigma^2(s)}\right) \tag{7}$$

The weight model-functions $w(s), \mu(s)$ and $\sigma(s)$ are parametrized and these parameters are collectively denoted by θ. We examined polynomial forms, i.e:

$$w(s) = \sum_{j=0}^{L_w} w_j s^j,\ \mu(s) = \sum_{j=0}^{L_\mu} \mu_j s^j \text{, and } \ \sigma(s) = \sum_{j=0}^{L_\sigma} \sigma_j s^j . \tag{8}$$

Note that $\mu_j\ \forall j = 0, \cdots, L_\mu$ and $\mu(s)$ are vectors in R^n. Therefore, the set of adjustable parameters becomes:

$$\theta = \{\{w_j\}_{j=0}^{L_w},\ \{\mu_{ij}\}_{i=1,j=0}^{n,\ L_\mu},\ \{\sigma_j\}_{j=0}^{L_\sigma}\} \tag{9}$$

with a total parameter number given by:

$$N_{var}^{FW} = (1+L_w) + n(L_\mu + 1) + (L_\sigma + 1) = L_w + nL_\mu + L_\sigma + n + 2 \tag{10}$$

The mean squared deviation corresponding to Eq. (3) is given by:

$$E_{[T]}(\theta) \stackrel{\text{def}}{=} \frac{1}{\#T} \sum_{\mathbf{x}^{(l)}, f^{(l)} \in T} \left(N_{FW}(\mathbf{x}^{(l)};\theta) - f^{(l)}\right)^2 \tag{11}$$

The above quantity serves as the objective function for the optimization procedure. Since $E_{[T]}(\theta)$ is continuously differentiable, Newton or Quasi-Newton methods are appropriate. Also, since such an objective function is expected to possess a multitude of local minima, a global procedure is necessary as usual [25].

2.2. Optimization strategy

Estimating the model parameters is accomplished by minimizing the mean squared error $E_{[T]}(\theta)$. Since this objective is multimodal, a global optimization technique known as "multistart" has been employed. This is a two-phase method, consisting of an exploratory global phase and a subsequent local minimum-seeking phase. In our case, the search during the local phase is performed via a Quasi-Newton approach with the BFGS update [26]. Since the network is expressed in a closed form, second derivatives may also be computed if a Newton method is to be preferred instead. In Multistart, a point θ is sampled uniformly from within the feasible region, $\theta \in S$, and subsequently a local search $\mathcal{L}$, is started from it leading to a local minimum $\widehat{\theta} = \mathcal{L}(\theta)$. If $\widehat{\theta}$ is a minimum found for the first time, it is stored, otherwise, it is rejected. The cycle goes on until a stopping rule [27] instructs termination.

3. Experimental results

We have performed a series of numerical experiments in order to test the performance of FWNN, and compare it to that of its predecessors, namely the sigmoid MLP and Gaussian RBF networks Both "homemade", as well as established benchmarks from the literature, have been employed. The homemade datasets were constructed by evaluating known functions at a number of selected points. Each set is divided into a training set and a test set. The training sets have been further manipulated by adding noise to the target values, in order to simulate the effect of "corruption" due to measurement errors, while the test sets have been left "clean", i.e. without any noise addition. The training of the FWNN is performed simply by minimizing the mean squared deviation given by Eq.(11), while for the MLP and RBF networks we minimize the corresponding "error" given in Eq.(3) with the addition of a penalty term, to facilitate the necessary "weight decay" regularization. Without regularization the MLP and RBF networks are susceptible to over-training. Since the mission of the network is to filter out the noise and reveal the underlying generating function, we have the luxury, for the homemade datasets, to train using noise-corrupted data, and evaluate the test error over clean data.

In the experiments, 10-fold cross-validation was adopted for the training. To quantify and evaluate network performance, the "Normalized Mean Squared Error" (NMSE) over the test set $S - T$, was used, namely:

$$NMSE = \frac{\sum\limits_{x^{(l)}, f^{(l)} \in S-T} \left(N_{RBF}(\mathbf{x}^{(l)}; \widehat{\theta}) - f^{(l)} \right)^2}{\sum\limits_{x^{(l)}, f^{(l)} \in S-T} \left(f^{(l)} \right)^2} \times 100 \,. \tag{12}$$

Note that the above metric is proper for comparisons, being almost insensitive to data scaling.

3.1. Experiments using artificial data

We conducted experiments with one- and two-dimensional data. Four functions were employed in one dimension and three in two dimensions. Graphical representations may be found in figures 1(a-d) and 2(a-c) correspondingly.

3.1.1. One-dimensional experiments For each function $f(x)$, an interval $[a, b]$ was chosen for the range of x values as shown in figures 1(a-d). The training set T, contains $L = 100$ equidistant points, determined by:

$$x_i = a + (i-1)\frac{b-a}{L-1}, \ \forall i = 1, 2, \cdots, L \tag{13}$$

The associated target values were then calculated as:

$$f_i = f(x_i) + \eta_i$$

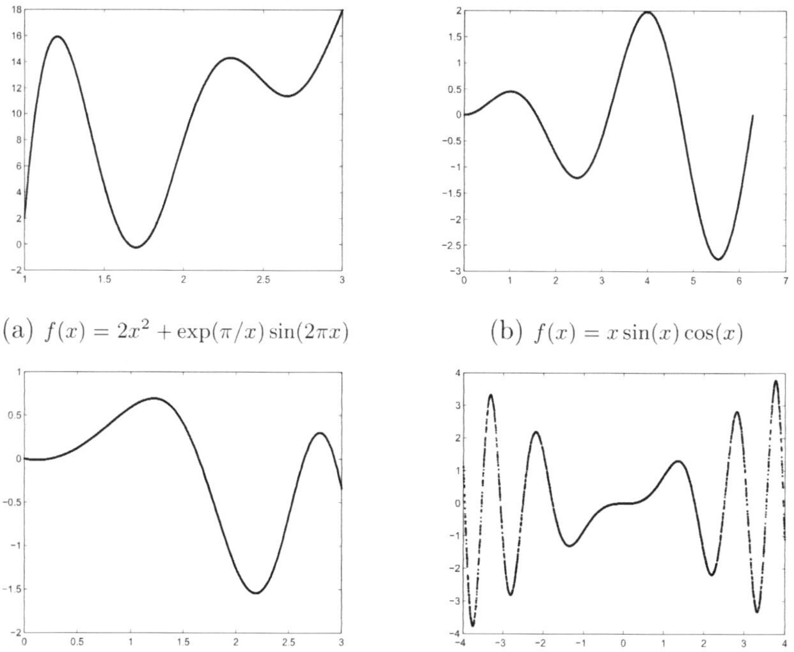

(a) $f(x) = 2x^2 + \exp(\pi/x)\sin(2\pi x)$ (b) $f(x) = x\sin(x)\cos(x)$

(c) $f(x) = \sin(x^2) - 0.25x$ (d) $f(x) = x\sin(x^2)$

Figure 1. Generating functions used for creating the $1d$ datasets. In each case 100 training and 1000 testing points were used.

with η_i being white Gaussian noise. Three levels of Signal-to-Noise-Ratio (SNR) were considered: no noise ($0\ dB$), medium ($-5\ dB$) and high noise ($-10\ dB$). Similarly the test set was created by picking $L = 1000$ equidistant points using eq. (13). No noise was added to the corresponding target values.

3.1.2. Two-dimensional experiments For each function $f(x_1, x_2)$, intervals $[a_1, b_1] \otimes [a_2, b_2]$ were chosen for the range of $x_1 \otimes x_2$ as shown in figures 2(a-c). For the training set, uniform grids of 15 points in each direction were used, resulting to a total of 225 training points. Again, we have added noise at three levels to the target values. Finally, for the test set, uniform grids of 100 points in each direction were used, resulting to a total of 10000 testing points (without any noise).

3.1.3. Procedural details FWNN, MLP and RBF networks have been investigated using the same datasets. For each noise level, 50 independent runs were executed and the corresponding mean NMSE value and its standard deviation are reported for both the training and the test sets. For the FWNN we used throughout the following values: $L_w = 5$, $L_\mu = 1$ and $L_\sigma = 1$, i.e. a fifth order polynomial for $w(s)$ and first order polynomials for $\mu(s)$ and $\sigma(s)$, corresponding to a total of $2n + 8$ parameters. For the case of MLP and RBF networks, four architectures, differing in the number of nodes, were used. In particular the instances of $K = 5, 10, 20$ and 30 nodes have been investigated, corresponding to $K(n + 2) + 1$ parameters.

In table 1, the results for the mean NMSE and its standard deviation, estimated over fifty independent runs, are laid out for the $1d$-examples. Accordingly, the results for the $2d$-examples

IOP Conf. Series: Journal of Physics: Conf. Series **915** (2017) 012006 doi:10.1088/1742-6596/915/1/012006

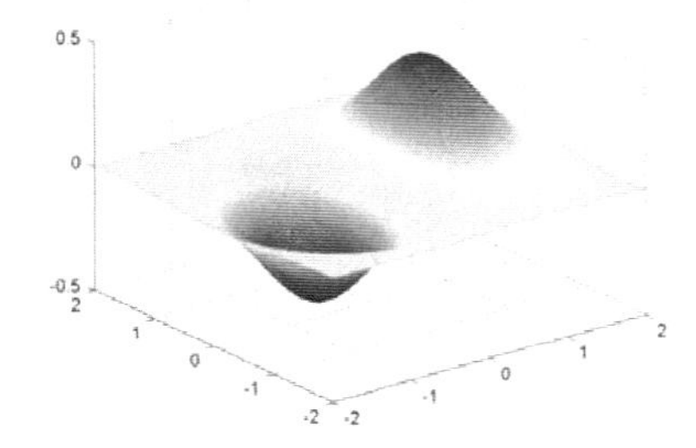

(a) $f(x_1, x_2) = x_1 \exp(-(x_1^2 + x_2^2))$

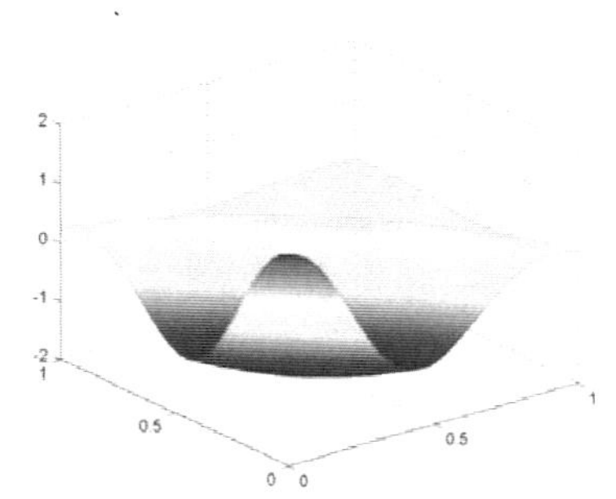

(b) $f(x_1, x_2) = \frac{\pi}{2} \exp(-2(x_1^2 + x_2^2)) \cos(2\pi(x_1 + x_2))$

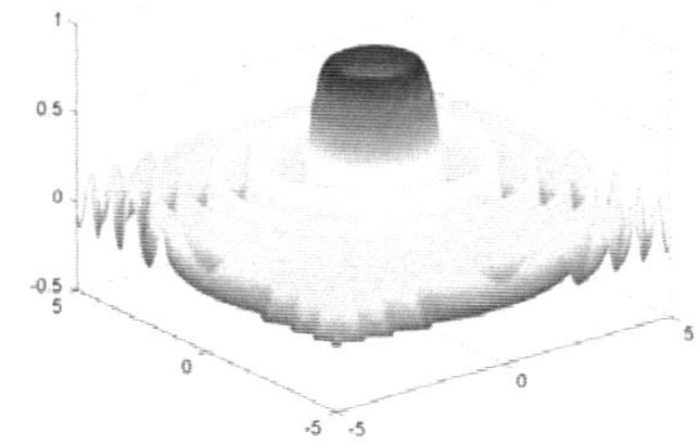

(c) $f(x_1, x_2) = \frac{\sin(x_1^2 + x_2^2)}{\sqrt{x_1^2 + x_2^2}}$

Figure 2. Generating functions used for creating the $2d$ datasets. In each case 225 training and 10000 testing points were used.

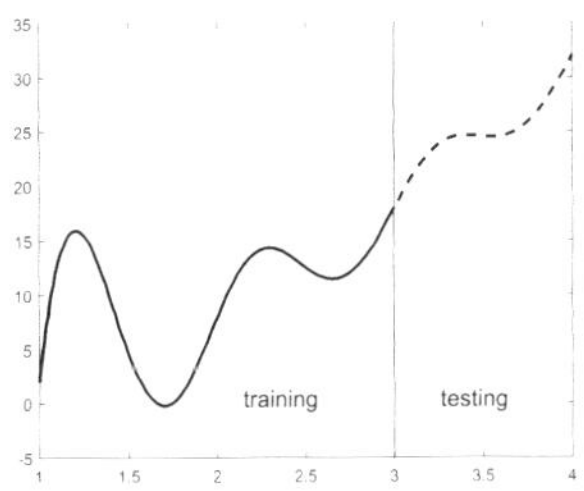

Dataset of figure 1a

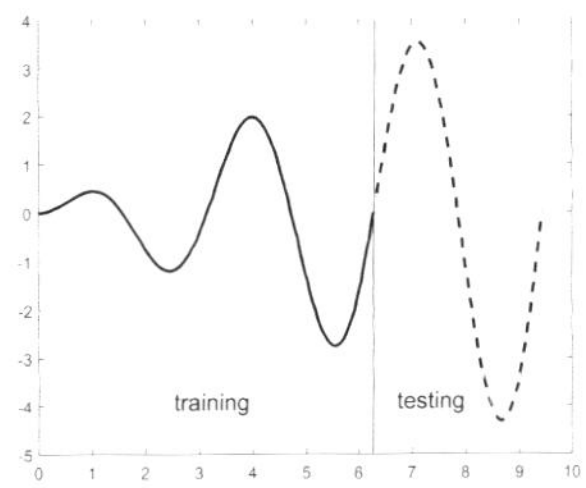

Dataset of figure 1b

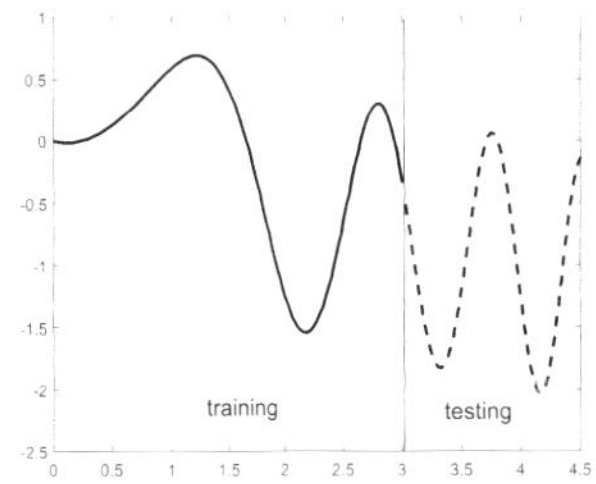

Dataset of figure 1c

Figure 3. The 3 datasets were created using three first generating functions shown in figure 1. They contain 150 equidistant points used for evaluating the extrapolation capabilities of the networks. The training sets (continuous lines) contain 100 points and the extrapolation test sets (dotted lines) 50 points.

are listed in table 2. All networks perform similarly for the $1d$ case, however FWNN has a clear edge in the case of highly corrupted data. This advantage becomes even more evident in the $2d$ case, where FWNN employs only 12 parameters to complete the task, yielding at the same time a superior level of generalization.

Table 1. NMSE mean and std comparison, for the 1*d* datasets related to figures 1(a-d), calculated over 50 independent experiments. The number in brackets is the number of parameters.

Network Architecture		*without noise*		*medium noise*		*high noise*	
		train	*test*	*train*	*test*	*train*	*test*
dataset 1(a)							
FWNN (10)		0.00 ± 0.00	$\mathbf{0.00 \pm 0.00}$	5.57 ± 0.81	0.89 ± 0.58	16.95 ± 1.01	$\mathbf{1.47 \pm 0.59}$
NN_5	MLP	0.00 ± 0.00	$\mathbf{0.00 \pm 0.00}$	5.98 ± 0.92	$\mathbf{0.86 \pm 0.45}$	15.88 ± 2.54	2.59 ± 0.83
(16)	RBF	0.15 ± 0.11	0.12 ± 0.08	6.45 ± 0.92	0.95 ± 0.54	17.56 ± 2.70	3.46 ± 2.33
NN_{10}	MLP	0.00 ± 0.00	0.00 ± 0.00	6.46 ± 0.79	0.93 ± 0.39	15.88 ± 3.67	2.45 ± 1.27
(31)	RBF	0.00 ± 0.00	$\mathbf{0.00 \pm 0.00}$	6.01 ± 1.11	1.71 ± 1.51	13.87 ± 1.86	4.19 ± 1.30
NN_{20}	MLP	0.00 ± 0.00	0.00 ± 0.00	5.66 ± 0.65	1.46 ± 1.19	15.67 ± 1.53	2.61 ± 0.83
(61)	RBF	0.00 ± 0.00	0.00 ± 0.00	5.13 ± 0.98	1.80 ± 1.72	11.70 ± 2.89	8.29 ± 6.24
NN_{30}	MLP	0.00 ± 0.00	0.00 ± 0.00	5.79 ± 1.47	1.75 ± 1.22	14.25 ± 2.66	4.46 ± 2.78
(91)	RBF	0.00 ± 0.00	0.00 ± 0.00	5.67 ± 1.10	1.29 ± 0.68	16.27 ± 3.17	2.90 ± 1.11
dataset 1(b)							
FWNN (10)		0.00 ± 0.00	$\mathbf{0.00 \pm 0.00}$	12.53 ± 1.61	$\mathbf{2.07 \pm 1.25}$	25.94 ± 3.67	$\mathbf{4.23 \pm 1.64}$
NN_5	MLP	0.00 ± 0.00	$\mathbf{0.00 \pm 0.00}$	12.40 ± 1.67	2.50 ± 0.98	28.04 ± 7.16	5.49 ± 3.37
(16)	RBF	0.18 ± 0.11	0.27 ± 0.17	12.59 ± 2.30	2.60 ± 1.21	24.19 ± 3.48	5.40 ± 2.30
NN_{10}	MLP	0.00 ± 0.00	0.00 ± 0.00	12.37 ± 1.67	2.13 ± 1.17	25.00 ± 2.57	4.40 ± 1.22
(31)	RBF	0.00 ± 0.00	$\mathbf{0.00 \pm 0.00}$	10.84 ± 3.04	2.69 ± 1.38	25.32 ± 4.22	7.86 ± 4.55
NN_{20}	MLP	0.00 ± 0.00	0.00 ± 0.00	11.49 ± 0.91	3.19 ± 1.46	26.48 ± 4.57	6.74 ± 2.67
(61)	RBF	0.00 ± 0.00	0.00 ± 0.00	7.80 ± 1.69	5.99 ± 1.65	24.30 ± 3.83	6.73 ± 2.44
NN_{30}	MLP	0.00 ± 0.00	0.00 ± 0.00	11.05 ± 2.06	5.66 ± 2.87	23.80 ± 3.75	10.88 ± 5.15
(91)	RBF	0.00 ± 0.00	0.00 ± 0.00	9.29 ± 2.39	5.34 ± 2.28	21.96 ± 3.98	8.85 ± 3.49
dataset 1(c)							
FWNN (10)		0.00 ± 0.00	$\mathbf{0.00 \pm 0.00}$	32.52 ± 3.56	$\mathbf{3.52 \pm 2.09}$	54.75 ± 2.84	$\mathbf{11.11 \pm 3.51}$
NN_5	MLP	0.00 ± 0.00	$\mathbf{0.00 \pm 0.00}$	33.20 ± 5.42	7.50 ± 3.05	57.74 ± 7.01	15.20 ± 6.71
(16)	RBF	0.08 ± 0.06	0.10 ± 0.10	32.40 ± 3.75	6.72 ± 1.12	50.93 ± 4.72	15.60 ± 11.58
NN_{10}	MLP	0.00 ± 0.00	0.00 ± 0.00	32.85 ± 4.65	5.55 ± 3.60	55.08 ± 7.36	16.02 ± 8.98
(31)	RBF	0.00 ± 0.00	$\mathbf{0.00 \pm 0.00}$	26.58 ± 4.53	9.59 ± 3.78	55.41 ± 6.89	16.13 ± 9.00
NN_{20}	MLP	0.00 ± 0.00	0.00 ± 0.00	29.74 ± 5.84	9.88 ± 6.02	53.98 ± 8.19	16.37 ± 8.53
(61)	RBF	0.00 ± 0.00	0.00 ± 0.00	21.97 ± 2.60	17.66 ± 7.50	51.80 ± 6.66	25.50 ± 14.95
NN_{30}	MLP	0.00 ± 0.00	0.00 ± 0.00	27.13 ± 4.94	13.96 ± 10.68	48.00 ± 6.39	28.45 ± 16.78
(91)	RBF	0.00 ± 0.00	0.00 ± 0.00	15.54 ± 2.19	24.98 ± 7.40	47.76 ± 6.39	22.27 ± 15.76
dataset 1(d)							
FWNN (10)		0.00 ± 0.00	$\mathbf{0.00 \pm 0.00}$	7.92 ± 1.59	$\mathbf{2.77 \pm 0.93}$	15.31 ± 2.54	$\mathbf{6.80 \pm 0.53}$
NN_5	MLP	95.08 ± 5.23	95.12 ± 5.23	99.03 ± 3.38	99.10 ± 4.29	98.55 ± 1.66	100.02 ± 1.76
(16)	RBF	60.50 ± 12.49	59.22 ± 13.19	61.32 ± 15.14	57.54 ± 16.47	56.35 ± 9.84	52.69 ± 11.09
NN_{10}	MLP	0.05 ± 0.05	0.05 ± 0.02	7.21 ± 0.97	3.83 ± 1.94	17.33 ± 4.21	9.94 ± 6.02
(31)	RBF	0.51 ± 1.58	0.32 ± 0.99	13.22 ± 6.95	7.46 ± 5.15	19.82 ± 5.31	11.76 ± 5.54
NN_{20}	MLP	0.00 ± 0.00	$\mathbf{0.00 \pm 0.00}$	7.06 ± 1.37	4.04 ± 1.35	16.10 ± 2.98	8.50 ± 2.80
(61)	RBF	0.00 ± 0.00	0.00 ± 0.00	8.60 ± 1.92	5.40 ± 1.82	18.96 ± 8.45	11.22 ± 11.19
NN_{30}	MLP	0.00 ± 0.00	0.00 ± 0.00	5.65 ± 1.31	4.66 ± 2.15	15.16 ± 3.14	12.15 ± 6.11
(91)	RBF	0.00 ± 0.00	0.00 ± 0.00	7.22 ± 2.00	3.29 ± 0.61	14.58 ± 3.31	9.08 ± 3.15

Table 2. NMSE mean and std comparison, for the $2d$ datasets related to figures 2(a-c), calculated over 50 independent experiments. The number in brackets is the number of parameters.

Network Architecture		*without noise*		*medium noise*		*high noise*	
		train	*test*	*train*	*test*	*train*	*test*
dataset 2(a)							
FWNN (12)		0.00 ± 0.00	$\mathbf{0.00 \pm 0.00}$	46.59 ± 3.04	$\mathbf{2.89 \pm 0.83}$	71.73 ± 2.08	$\mathbf{4.66 \pm 2.19}$
NN_5	MLP	4.30 ± 0.24	3.82 ± 0.26	51.97 ± 3.94	10.57 ± 2.00	74.88 ± 3.88	20.43 ± 11.11
(21)	RBF	0.00 ± 0.00	$\mathbf{0.00 \pm 0.00}$	48.63 ± 2.66	3.36 ± 1.16	71.68 ± 2.50	7.42 ± 3.14
NN_{10}	MLP	0.11 ± 0.00	0.09 ± 0.00	49.82 ± 3.41	8.08 ± 1.60	73.13 ± 4.26	15.83 ± 5.00
(41)	RBF	0.00 ± 0.00	0.00 ± 0.00	47.07 ± 2.52	5.37 ± 2.05	71.27 ± 2.76	9.92 ± 5.31
NN_{20}	MLP	0.02 ± 0.00	0.02 ± 0.00	48.72 ± 3.23	7.43 ± 1.92	74.71 ± 3.35	16.65 ± 3.97
(81)	RBF	0.00 ± 0.00	0.00 ± 0.00	45.33 ± 3.21	6.27 ± 2.41	68.74 ± 2.99	13.25 ± 5.96
NN_{30}	MLP	0.02 ± 0.00	0.01 ± 0.00	45.98 ± 3.10	7.11 ± 1.66	71.73 ± 5.28	14.84 ± 3.78
(121)	RBF	0.00 ± 0.00	0.00 ± 0.00	46.19 ± 4.18	6.79 ± 3.26	69.39 ± 2.93	12.11 ± 6.35
dataset 2(b)							
FWNN (12)		0.00 ± 0.00	$\mathbf{0.00 \pm 0.00}$	51.23 ± 4.69	$\mathbf{5.14 \pm 0.71}$	73.91 ± 4.35	$\mathbf{9.53 \pm 2.59}$
NN_5	MLP	0.19 ± 0.05	0.17 ± 0.05	51.16 ± 0.66	5.42 ± 2.14	74.34 ± 8.98	19.27 ± 26.57
(21)	RBF	3.19 ± 1.69	2.90 ± 1.58	51.75 ± 3.94	8.79 ± 4.61	72.21 ± 2.67	15.11 ± 3.49
NN_{10}	MLP	0.01 ± 0.01	0.01 ± 0.00	50.50 ± 2.93	6.60 ± 2.63	70.33 ± 2.87	12.88 ± 6.68
(41)	RBF	0.03 ± 0.01	0.03 ± 0.01	47.61 ± 2.46	6.15 ± 0.76	69.41 ± 2.66	13.37 ± 5.14
NN_{20}	MLP	0.01 ± 0.00	0.00 ± 0.00	49.82 ± 2.80	7.17 ± 2.62	71.81 ± 2.80	11.03 ± 2.81
(81)	RBF	0.00 ± 0.00	0.00 ± 0.00	45.87 ± 2.77	11.12 ± 3.61	68.07 ± 2.53	16.04 ± 4.35
NN_{30}	MLP	0.01 ± 0.00	0.00 ± 0.00	48.64 ± 2.13	6.09 ± 2.09	70.67 ± 2.81	11.57 ± 4.04
(121)	RBF	0.00 ± 0.00	0.00 ± 0.00	45.45 ± 2.83	11.05 ± 3.32	66.68 ± 3.76	24.91 ± 8.30
dataset 2(c)							
FWNN (12)		22.35 ± 1.42	$\mathbf{20.95 \pm 1.18}$	64.43 ± 2.90	$\mathbf{26.76 \pm 2.05}$	84.27 ± 2.85	$\mathbf{63.93 \pm 18.58}$
NN_5	MLP	96.44 ± 2.32	96.32 ± 2.60	96.95 ± 1.80	96.72 ± 1.79	98.51 ± 1.86	98.23 ± 1.72
(21)	RBF	65.32 ± 4.34	63.69 ± 4.42	83.19 ± 2.17	66.98 ± 4.11	89.61 ± 2.75	72.10 ± 3.48
NN_{10}	MLP	87.89 ± 9.93	87.26 ± 10.36	98.44 ± 1.34	97.82 ± 1.50	97.68 ± 2.46	98.70 ± 2.15
(41)	RBF	62.23 ± 5.85	$60.59 \pm 5.88)$	78.88 ± 2.52	$67.88 \pm 2.89)$	85.54 ± 3.77	81.04 ± 11.50
NN_{20}	MLP	89.16 ± 9.73	88.85 ± 10.03	97.17 ± 2.84	96.86 ± 2.74	97.00 ± 3.22	98.43 ± 1.63
(81)	RBF	47.15 ± 3.15	45.25 ± 3.07	69.85 ± 4.77	67.12 ± 10.16	79.51 ± 6.39	94.86 ± 13.43
NN_{30}	MLP	71.89 ± 2.69	70.90 ± 2.88	95.58 ± 3.54	99.46 ± 3.69	94.79 ± 2.87	91.89 ± 4.07
(121)	RBF	47.15 ± 3.15	40.07 ± 2.22	64.67 ± 4.46	68.29 ± 4.51	74.41 ± 5.84	111.65 ± 20.54

3.2. Extrapolation

FWNN is evaluated as trustworthy as far as interpolation is concerned. It would be quite interesting to explore its extrapolation capability as well. Although this was not the main goal of the present article, we were tempted to examine a few one-dimensional cases to obtain an indication of how its extrapolation capability compares to that of the traditional MLP and RBF networks. We describe briefly the methodology used for the evaluation of the extrapolation performance. We construct a grid of 150 equidistant points. The first 100 points are used for training the networks. The remaining 50 points, $x_1, x_2, \cdots, x_{50}$, are used for evaluating the extrapolation quality. Three such datasets were used as depicted in figure 3.

Let $f(x)$ and $N(x)$ denote the generating function and the corresponding trained network; then we may consider as a measure for the network's extrapolation capability, the relative

Table 3. Comparison of the extrapolation index J, for the 3 datasets.

Network Architecture	*Deviation bound d*				
	0.05	0.10	0.15	0.20	0.25
dataset of Fig. 1a					
FWNN	25 ± 1	38 ± 2	50 ± 0	50 ± 0	50 ± 0
MLP 10 nodes	15 ± 2	25 ± 3	31 ± 4	34 ± 5	50 ± 0
RBF 10 nodes	15 ± 2	19 ± 2	22 ± 3	27 ± 5	29 ± 4
dataset of Fig. 1b					
FWNN	24 ± 3	25 ± 3	35 ± 2	37 ± 2	38 ± 2
MLP 10 nodes	12 ± 2	14 ± 2	15 ± 3	16 ± 1	16 ± 1
RBF 10 nodes	14 ± 2	16 ± 2	17 ± 2	18 ± 1	19 ± 1
dataset of Fig. 1c					
FWNN	18 ± 1	20 ± 1	21 ± 1	21 ± 1	21 ± 1
MLP 10 nodes	6 ± 2	8 ± 1	9 ± 1	10 ± 1	11 ± 1
RBF 10 nodes	11 ± 1	13 ± 2	14 ± 2	15 ± 1	16 ± 1

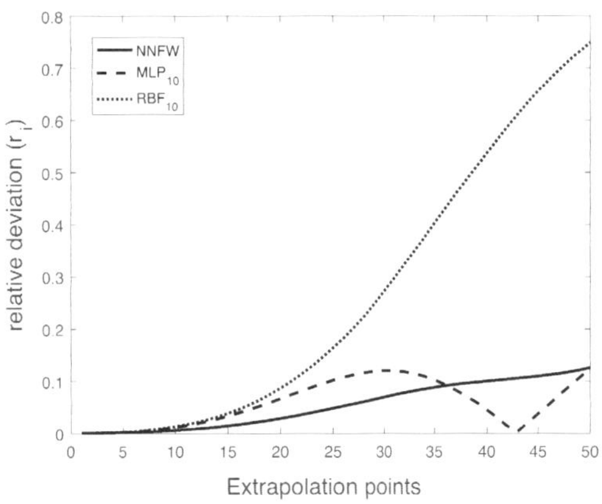

(a) Dataset: figure 1a

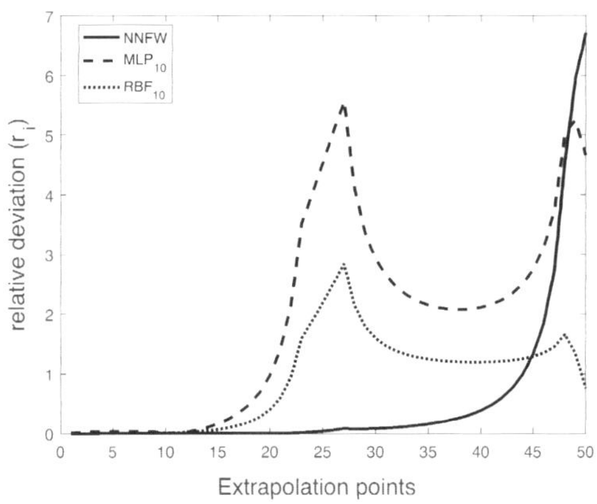

(b) Dataset: figure 1b

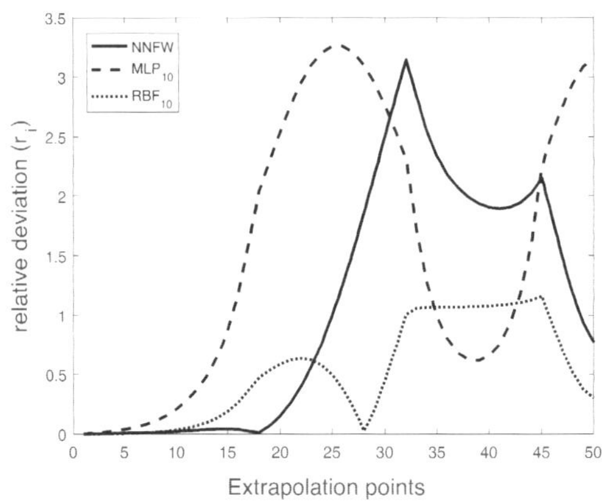

(c) Dataset: figure 1c

Figure 4. FWNN, MLP and RBF relative-deviations r_i, defined in eq. (14), evaluated at 50 equidistant points in the extrapolation domain, for the quoted datasets.

deviation r_i at point x_i given by:

$$r_i \equiv \frac{|f(x_i) - N(x_i)|}{\max\{1, |f(x_i)|\}} \tag{14}$$

Let J be the maximum number of consecutive points x_i starting from x_1 and satisfying $r_i < d$, where $d \in [0, 0.25]$, is a bound for the relative deviation. Namely J is determined so that:

$$r_i < d, \ \forall\ i \leq J \text{ and } \ r_{J+1} > d$$

Given a value for d, the best network for extrapolation is the one corresponding to the highest J. We list our findings for $d = 0.05, 0.10, 0.15, 0.20, 0.25$ in table 3. The mean values and the standard deviations of the J-index, have resulted from the conduction of 50 independent experiments. In figure 4(a-c) we compare the relative deviations r_i, $\forall i = 1, \cdots, 50$, of the FWNN, MLP and RBF networks, for three quoted datasets. In all cases the proposed FWNN, displays superior extrapolation performance.

4. Discussion and Conclusions

In this study we have proposed a new type of neural network, the FWNN, in which all of its parameters are functions of a continuous variable and not of a discrete index. This maybe interpreted as a neural network with an infinite number of hidden nodes and significantly restricted number of model parameters. In the conducted experiments on a suite of benchmark datasets, FWNN achieved superior generalization performance compared to its peers, namely the conventional MLP and RBF networks. In addition, FWNN is robust and effective even on difficult problems.

The FWNN has interesting properties. The important issue of generalization is being served superbly. The number of necessary parameters is rather limited, resisting so over-training. The positions of the Gaussian centers are determined by the arc $\mu(s), \forall s \in [-1, 1]$ and the corresponding spherical spreads by $\sigma(s)$. In the present article we have used linear forms, hence the arc is a straight line segment joining the two end points $\mu(-1)$ and $\mu(+1)$ in R^n. The spreads are linearly increasing or decreasing with s, depending on the sign of σ_1. Although this seems to be a severe constraint, it has not degraded the network's performance. This may be understood noting that the infinite number of nodes, renders the approximation of any function feasible, while the existence of the constraint, deters over-training by substantially restricting the number of adjustable parameters. If in some cases the linear model for $\mu(s)$ or $\sigma(s)$ imposes an overly strict constraint, it can be replaced by a more flexible quadratic, or cubic model, at the expense of some extra parameters. The Gaussian centers will then be lying on a parabolic, or a cubic arc correspondingly and the spreads will exhibit a higher level of adaptivity.

References

[1] Cybenko G 1989 *Mathematics of Control, Signals, and Systems* 303–314
[2] Hornik K 1991 *Neural Networks* 251–257
[3] Ripley B D 1996 *Pattern Recognition and Neural Networks* (Campridge University Press)
[4] Bishop C 2006 *Pattern Recognition and Machine Learning* (Springer)
[5] Lagaris I E, Likas A and Fotiadis D I 1997 *Computer Physics Communications* 1–14
[6] Lagaris I E, Likas A and Fotiadis D I 1998 *IEEE Transactions on Neural Networks* 987–1000
[7] Lagaris I E, Likas A and Papageorgiou D G 2000 *IEEE Transactions on Neural Networks* **11** 1041–1049
[8] Wang J Z, Wang J J, Zhang Z G and Guo S P 2011 *Expert Systems with Applications* **38** 14346–14355
[9] Bartlett P 1998 *IEEE Transactions on Information Theory* **44** 525–536
[10] MacKay D J C 1992 *Neural Computation* **4** 415–447
[11] Sietsma J and Dow R J 1991 *Neural Networks* **4** 67–79
[12] Murphy K 2012 *Machine Learning: A Probabilistic Perspective* (MIT Press)
[13] Tibshirani R 1996 *Journal of the Royal Statistical Society B* **58** 267–288

IOP Conf. Series: Journal of Physics: Conf. Series **915** (2017) 012006 doi:10.1088/1742-6596/915/1/012006

[14] Meyer D, Leisch F and Hornik K 2003 *Neurocomputing* **55** 169–186
[15] Tipping M 2001 *Journal of Machine Learning Research* **1** 211–244
[16] Seeger M 2008 *Journal of Machine Learning Research* **9** 759–813
[17] Tzikas D, Likas A and Galatsanos N 2009 *IEEE Trans. on Neural Networks* **20** 926–937
[18] Blekas K and Likas A 2014 *Knowl. Inf. Syst.* **39** 241–264
[19] Srivastav N, Hinton G, Krizhevsky A, Sutskever I and Salakhutdinov R 2014 *Journal of Machine Learning Research* **15** 1929–1958
[20] Roux N and Bengio Y 2007 *Eleventh International Conference on Artificial and Statistics (AISTATS)* pp 404–411
[21] Powell M 1985 *IMA conference on "Algorithms for the Approximation of Functions and Data* (RMCS Shrivenham)
[22] Broomhead D S and Lowe D 1988 *Complex Systems* **2** 321–355
[23] Geman S, Bienenstock E and Doursat R 1992 *Neural Computation* **4** 1–58
[24] Krogh A and Hertz J 1992 *Advances in Neural Information Processing Systems* vol 4 pp 950–957
[25] Voglis C and Lagaris I 2006 *Neural, Parallel & Scientific Computations* **14** 231–240
[26] Fletcher R 1970 *Computer Journal* **13** 317–322
[27] Lagaris I E and Tsoulos I G 2008 *Applied Mathematics and Computation* **197** 622–632

IOP Conf. Series: Journal of Physics: Conf. Series **915** (2017) 012007 doi:10.1088/1742-6596/915/1/012007

Jost function method approach for study of unstable nuclei

Hiroshi Masui

Information Processing Center, Kitami Institute of Technology, 165 Koen-Cho, Kitami, 090-8507, Japan

E-mail: hgmasui@mail.kitami-it.ac.jp

Abstract. We have developed the formalism of the Jost function method (JFM) to study unstable nuclei. We apply the JFM to the calculations, the partial decay widths in the coupled-channel problem, anti-bound (virtual) states, non-local potentials and Lagrange-mesh method. The results are compared with other methods, and we show the JFM formalism is useful to study the unstable nuclei.

1. Introduction

The component of the unbound states becomes important in the field of quantum systems where open channels are included in the system, e.g. study of unstable nuclei. From the aspect of the few-body treatment, the position of the S-matrix pole is a crucial key to characterize the system. The pole position can be obtained very precisely by using the Jost function method (JFM). A practical recipe for the numerical treatment of JFM was given by Sofianos and Rakityanskiy [1]. We also have developed the formalism of the Jost function method (JFM) to study the partial decay widths in coupled-channel systems [2], virtual (anti-bound) states [3], non-local kernels in the Hamiltonian [4] and application to the Lagrange-mesh formalism [5].

We have developed the applications of JFM for studying the unstable nuclei. First, we have apply the partial decay widths in a coupled-channel system can be determined by JFM under the assumption that the sum of the partial widths becomes the total width [2]. We demonstrate how accurately we can determine the partial decay widths even for broad resonant cases and show that two different schemes to define the widths give exactly the same result using the JFM approach.

Next important progress on the study of unstable nuclei is the discussion of the virtual state pole within the other physical observable, i.e. scattering phase-shift and scattering length. We compare the two systems, ^{5}He and ^{10}Li [3], which are considered to be the subsystem of halo nuclei: ^{6}He and ^{11}Li. To understand the structure of ^{11}Li, the position of the pole in s-wave state of ^{10}Li is important. Because ^{11}Li is a p-shell nucleus in the shell model point of view, and the valence neutrons are in the $0p_{1/2}$-orbit with respect to the ^{9}Li core. Nevertheless, experiments indicate the strong s-wave component for the low-lying state of ^{10}Li (^{9}Li+n) system, and no s-wave bound states are observed. Hence, such the s-state can be considered as a virtual (anti-bound) state. For the theoretical approach to study of ^{11}Li, we need to determine the potential strength of the ^{9}Li+n system. The position of the S-matrix pole can be investigated by JFM even for the virtual states, and we discuss how the pole moves on the complex momentum plane

IOP Conf. Series: Journal of Physics: Conf. Series **915** (2017) 012007 doi:10.1088/1742-6596/915/1/012007

by changing the potential strength. The results are connected to other physical observable such as the phase shift and scattering length.

The remaining issues for the JFM approach is to include non-local potentials or kernels into the formalism [4]. We proposed a practical approach for solving an integro-differential equation of JFM. Using this formalism, the virtual state of ^{10}Li can be determined including the non-local Pauli projection operators under the orthogonality condition model.

Recently, we apply JFM to the Lagrange-mesh approach [5]. The prominent advantages of the Lagrange-mesh approach are follows. The mesh points of this method can be taken very small number compared to the conventional numerical integration for solving the differential equation, e.g. the Runge-Kutta method. The typical number of the mesh points is at most 100. Furthermore, the mesh points are defined as the zero-point of the Lagrange polynomials. Hence, the non-local potential which induces the integration in the differential equation can be reduced to a value to each mesh point.

2. Formalism of the Jost function method

We briefly review the Jost function method (JFM). Details are shown in Refs. [1, 2]

We consider the Schrödinger equation for the radial part of a one- or two-body system as follows:

$$\left[\frac{\partial^2}{\partial r^2} + k^2 - \frac{2\eta k}{r} - \frac{l(l+1)}{r^2}\right]\phi(r) = \frac{2\mu}{\hbar^2}V(r)\,\phi(r)\ , \tag{1}$$

where the radial part of the wave function $R(r)$ is replaced as $u(r) = rR(r)$. The homogeneous solution $H_l^{(\pm)}(z)$ of Eq. (1) becomes known functions, i.e. F_l and G_l; the regular and irregular Coulomb functions and j_l and n_l; spherical Bessel functions.

In order to solve Eq. (1), we introduce two unknown functions $\mathcal{F}^{(\pm)}(k,r)$ and define the regular solution as follows:

$$u(r) \equiv \frac{1}{2}\left[H_l^{(+)}(kr)\mathcal{F}^{(+)}(k,r) + H_l^{(-)}(kr)\mathcal{F}^{(-)}(k,r)\right]\ . \tag{2}$$

The functions $\mathcal{F}^{(\pm)}(k,r)$ become a constant for $r \to \infty$ that correspond to the Jost functions. Since we introduced two unknown functions $\mathcal{F}^{(\pm)}(k,r)$ for solving the equation, we can define a constraint to reduce the degrees of freedom. Thus we define the constraint as

$$H_l^{(+)}\left[\partial_r\mathcal{F}^{(+)}\right] + H_l^{(-)}\left[\partial_r\mathcal{F}^{(-)}\right] = 0\ . \tag{3}$$

Inserting Eq. (2) into Eq (1) and using the condition (3), the second-order differential equation (1) is reduced to the first-order one:

$$\frac{\partial\mathcal{F}^{(\pm)}(k,r)}{\partial r} = \pm\frac{\mu}{ik\hbar^2}H_l^{(\mp)}(kr)V(r)\left\{H_l^{(+)}(kr)\mathcal{F}^{(+)}(k,r) + H_l^{(-)}(kr)\mathcal{F}^{(-)}(k,r)\right\}\ . \tag{4}$$

This is the basic equation of the Jost function method, and we can solve the equation using ordinary numerical techniques such as the Runge-Kutta method.

For the boundary condition of the function $\mathcal{F}^{(\pm)}(k,r)$, we have

$$\lim_{r\to 0} F^{(\pm)}(k,r) = 1\ , \tag{5}$$

due to the boundary condition that the wave function is regular at the origin.

3. Applications of JFM to study of unstable nuclei

In this section, we show examples of the application of JFM to the study of unstable nuclei and related subjects.

3.1. Partial decay widths

We discuss the treatment of the partial decay widths of coupled-channel systems using a numerical example, the Noro-Taylor model [6]. The potential part of the Noro-Taylor model in the two-channel system is taken as

$$V_{nn'} = \lambda_{nn'} r^2 \exp(-r) + E_n \delta_{nn'} \;, \tag{6}$$

where the $\lambda_{nn'}$ are potential strengths

$$\lambda_{nn'} = \begin{pmatrix} -1.0 & -7.5 \\ -7.5 & 7.5 \end{pmatrix} , \tag{7}$$

and E_n are the threshold energies $E_1 = 0$ and $E_2 = 0.1$. The reduced masses are taken as $\hbar/\mu_n = 1$. Here all units are taken as a.u.

We calculate the resonant poles of the Noro-Taylor model using JFM and list the result in table 1. Different form the complex scaling method approach, JFM can give the complex energy of the poles even for very broad resonant states as shown in table 1.

Table 1. Resonant poles of the Noro-Taylor model [6] on the complex energy plane. All units are in a.u.

pole	E_r	Γ
1	4.768197	1.420192 $\times 10^{-3}$
2	7.241200	1.511912
3	8.171216	6.508332
4	8.440526	12.56299
5	8.072642	19.14563
6	7.123813	26.02534
7	5.641023	33.07014

Next, we discuss the calculational schemes for the partial decay widths using JFM. We compare the three different schemes for the partial decay widths, (i) the S-matrix scheme, (ii) the T-matrix scheme, and (iii) the current density scheme.

For a well-isolated resonant state, an S-matrix $S_{nn'}(E)$ is expressed in the Breit-Wigner form with a resonance energy E_r, a total width Γ, and partial decay widths to channels n and n' (Γ_n and $\Gamma_{n'}$, respectively) as

$$S_{nn'}(E) \equiv S_n^{\rm bg}(E)\,\delta_{nn'} - i\frac{\sqrt{\Gamma_n \cdot \Gamma_{n'}}}{E - E_r + i\Gamma/2} \;. \tag{8}$$

From Eq. (8), the ratio of the partial decay widths can be expressed using the ratio of the S-matrix as follows:

$$\frac{\Gamma_n}{\Gamma_{n'}} = \left| \frac{\lim\limits_{E\to E_{\rm res}} (E - E_{\rm res}) S_{nn}(E)}{\lim\limits_{E\to E_{\rm res}} (E - E_{\rm res}) S_{n'n'}(E)} \right| = \left| \frac{S_{nn}(E_{\rm res})}{S_{n'n'}(E_{\rm res})} \right| = \left| \frac{\sum_m \mathcal{F}^{(+)}_{nm}(E_{\rm res},\infty)\cdot \mathcal{G}_{mn}(E_{\rm res})}{\sum_m \mathcal{F}^{(+)}_{n'm}(E_{\rm res},\infty)\cdot \mathcal{G}_{mn'}(E_{\rm res})} \right| \;. \tag{9}$$

Here, we use the definition of the S-matrix using JFM as

$$\left(||\mathcal{F}^{(-)}(E,\infty)||^{-1} \right)_{mn} = \frac{1}{\det||\mathcal{F}^{(-)}(E,\infty)||}\, \mathcal{G}_{mn}(E) \;, \tag{10}$$

where $\|\mathcal{G}(E)\|$ is an adjoint matrix of $\|\mathcal{F}^{(-)}(E,\infty)\|$. In the above expression, the origin of the pole in the S-matrix comes from the root of $\det\|\mathcal{F}^{(-)}(E_{\rm res},\infty)\| = 0$. Therefore, we can eliminate the singularity and obtain the Eq. (9). In table 2, we compare the results of three different schemes for the calculation of the partial decay widths. We show the three schemes are equivalent each other and gives the same numerical result in a good accuracy.

Table 2. Comparison of results for the calculation with various methods. All units are in a.u.

method	Γ_1	Γ_2
Present: S-matrix	5.1116×10^{-5}	1.369076×10^{-3}
Present: T-matrix	5.1116×10^{-5}	1.369076×10^{-3}
Residue of S-matrix in Ref. [1]	5.1103×10^{-5}	1.368733×10^{-3}
Noro-Taylor [6]	5.9×10^{-5}	1.361×10^{-3}
T-matrix in Ref. [7]	5.1×10^{-5}	1.368×10^{-3}
Current density in Ref. [7]	5×10^{-5}	1.45×10^{-3}

3.2. Virtual states in ^{5}He and ^{10}Li

In this subsection, we discuss the pole position of the virtual states in ^{5}He and ^{10}Li [3].

First, we study the 1s-state of the ^{4}He+n system (^{5}He), which will provide important information on the halo structure of ^{6}He as a subsystem of the ^{4}He+n+n model. For this purpose, we use the so-called KKNN potential [8]. Using the complex scaling method, the resonant poles of p- and d-states can be obtained. On the other hand, due to a calculational difficulty for the virtual states, the position of the poles of the s-wave states have been not investigated. Therefore, we investigate the s-wave virtual pole using JFM, which has no difficulty for calculating the virtual states. From the analysis of the phase shift and position of the poles, we conclude that the 1s-state in the ^{4}He+n system cannot be a virtual state without changing the scattering phase shift drastically, and even if the system becomes the virtual state, the imaginary part becomes very large.

Next, we study the 1s-state in the ^{9}Li+n system (^{10}Li). We investigate the behavior of the 1s-state pole in the ^{9}Li+n system with changing the potential strength, of which we call "FP1". we use a parameter $(1+\delta)$ for the attractive part of the ^{9}Li-n central potential. The trajectories of the pole of the 1s-state are shown in figure 1. We take the range of the potential strength δ from -0.5 to 0.5. For $\delta = 0$, the 1s-state becomes a virtual state. The virtual state is realized in the range of the strength δ as $-0.26 < \delta < 0.07$.

We compare our result for the ^{10}Li with ones obtained by Thompson and Zhukov [9]. The result is shown in figure 2. As shown in figure 2, for the poles of P1 to P4, these positions are sensitive to the potential strength, since if we vary the strength with a very small value, the pole easily changes its position. The poles of P1 to P4 correspond our FP1 in the range of $0 < \delta < 0.07$.

3.3. Non-local operators

We consider the general case for an energy independent non-local potential $\Lambda(r,r')$ by taking

$$V(r) \Rightarrow V(r)\delta(r,r') + \Lambda(r,r') \ . \tag{11}$$

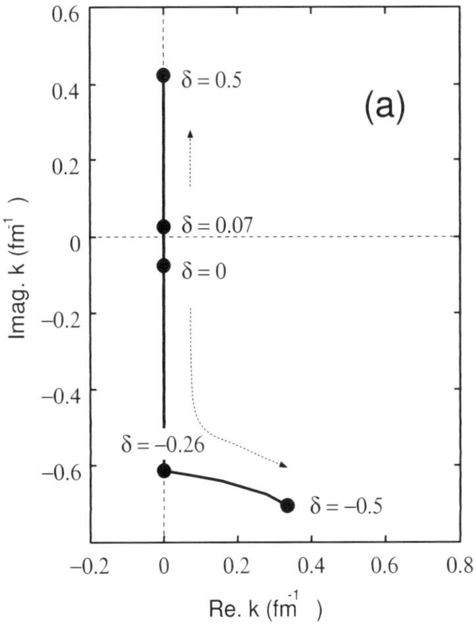

Figure 1. Pole behavior for ^{10}Li on the complex momentum plane.

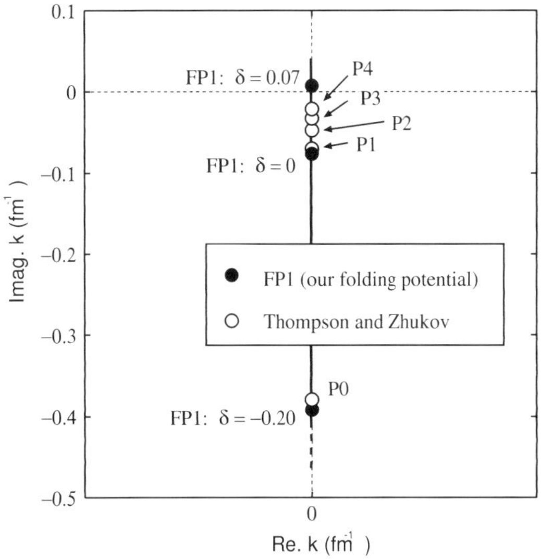

Figure 2. Comparison of the position of the poles with FP1 and those obtained in Ref. [20] (P0-P4).

Hence, the Schrödinger equation becomes the following form:

$$(E - \hat{H})|\phi\rangle = \hat{\Lambda}|\phi\rangle \ . \tag{12}$$

Using the JFM formalism, we obtain the integro-differential equation as

$$\begin{aligned}\frac{\partial \mathcal{F}^{(\pm)}_{nm}}{\partial r} &= \sum_{n'=1}^{N} \left[G^{(\pm,+)}_{nn'}(r)\mathcal{F}^{(+)}_{n'm}(k_n, r) + G^{(\pm,-)}_{nn'}(r)\mathcal{F}^{(-)}_{n'm}(k_n, r)\right] \\ &+ \sum_{n'=1}^{N} \int_0^\infty dr' \left[K^{(\pm,+)}_{nn'}(r,r')\mathcal{F}^{(+)}_{n'm}(k_{n'}, r') + K^{(\pm,-)}_{nn'}(r,r')\mathcal{F}^{(-)}_{n'm}(k_{n'}, r')\right] \ .\end{aligned} \tag{13}$$

Here, in order to simplify the expression of above equation, we define G and K as follows:

$$\left\{\begin{aligned} G^{(\pm,+)}_{nn'}(r) &\equiv \pm\frac{\mu_n}{ik_n\hbar} H^{(\mp)}_n(k_n r)\, V_{nn'}(r)\, H^{(+)}_{n'}(k'_n r) \ , \\ G^{(\pm,-)}_{nn'}(r) &\equiv \pm\frac{\mu_n}{ik_n\hbar} H^{(\mp)}_n(k_n r)\, V_{nn'}(r)\, H^{(-)}_{n'}(k'_n r) \ , \end{aligned}\right. \tag{14}$$

and

$$\left\{\begin{aligned} K^{(\pm,+)}_{nn'}(r,r') &\equiv \pm\frac{\mu_n}{ik_n\hbar} H^{(\mp)}_n(k_n r)\, \Lambda_{nn'}(r,r')\, H^{(+)}_{n'}(k'_n r') \ , \\ K^{(\pm,-)}_{nn'}(r,r') &\equiv \pm\frac{\mu_n}{ik_n\hbar} H^{(\mp)}_n(k_n r)\, \Lambda_{nn'}(r,r')\, H^{(-)}_{n'}(k'_n r') \ . \end{aligned}\right. \tag{15}$$

We introduce a matrix representation for the functions in Eq. (13) defined on the mesh points and obtain the equation as follows:

$$\boldsymbol{F}(\boldsymbol{r}) = \boldsymbol{F}(\boldsymbol{0}) + \boldsymbol{G}(\boldsymbol{r})\boldsymbol{F}(\boldsymbol{r}) + \boldsymbol{K}(\boldsymbol{r},\boldsymbol{r}')\boldsymbol{F}(\boldsymbol{r}') \ . \tag{16}$$

Finally, the matrix of the unknown function $\boldsymbol{F}(\boldsymbol{r})$ can be solved using the matrix inversion,

$$\boldsymbol{F}(\boldsymbol{r}) = \left[\mathbf{1} - \boldsymbol{G}(\boldsymbol{r}) - \boldsymbol{K}(\boldsymbol{r},\boldsymbol{r}')\right]^{-1} \boldsymbol{F}(\mathbf{0}) \ . \tag{17}$$

In order to confirm the validity of our approach, we compare the numerical result for the coupled-channel ^{9}Li+n system system. Results are shown in table 3. As shown in table 3, the JFM formalism with the matrix inversion on the mesh point can provide the accurate result as compared to the complex scaling method [10, 11, 12]. Different form the CSM approach, JFM can explicitly calculate the position of the virtual state pole on both the complex energy and momentum planes. This is a great advantage of the JFM approach in the study of unstable nuclei.

Table 3. Resonant and virtual states of the coupled-channel ^{9}Li+n system for the potential strength used in Ref. [13], $\delta = 0.245$. Here "b.s." indicates the bound state and "v.s." the virtual state.

state	CSM [13]	JFM-OCM
s-waves 2^-	-0.028 (b.s.)	-0.028 (b.s.)
1^-	—	-0.018 (v.s.)
p-wave 2^+	$0.71(\Gamma = 0.40)$	$0.73(\Gamma = 0.42)$
1^+	$0.42(\Gamma = 0.19)$	$0.42(\Gamma = 0.19)$

3.4. Lagrange-mesh method

Before we formulate the JFM on the Lagrange-mesh method, we briefly show the essential formalism of the method of Lagrange mesh. Details are shown in Refs. [14, 15, 16, 17, 18].

An integration of a function $g(x)$ in the interval $[a, b]$ can be approximated by using the Gauss quadrature,

$$\int_a^b g(x)dx \simeq \sum_{k=1}^{N} \lambda_k \, g(x_k) \ . \tag{18}$$

This quadrature can be defined a product of two functions as

$$g(x) \Rightarrow f_i(x)\, f_j(x) \ . \tag{19}$$

Here, the $f_i(x)$ are the Lagrange-basis functions, which satisfy the conditions at the mesh point x_j as follows:

$$f_i(x_j) = \lambda_i^{-1/2}\, \delta_{ij} \ . \tag{20}$$

These functions are orthonormal,

$$\int_a^b f_i(x)\, f_j(x)\, dx \ = \ \sum_{k=1}^{N} \lambda_k\, f_i(x_k)\, f_j(x_k) = \delta_{ij} \ . \tag{21}$$

Using the mesh points x_k and functions $f_i(x_k)$, we can formulate the matrix representation of JFM as follows:

$$\boldsymbol{F}_\infty - \boldsymbol{F}_0 \simeq \boldsymbol{A}_G\, \boldsymbol{F}_\infty + \tilde{\boldsymbol{B}}_{ex}^T \tilde{\boldsymbol{M}}^{-1} \tilde{\boldsymbol{B}} \boldsymbol{F}_\infty \ . \tag{22}$$

IOP Conf. Series: Journal of Physics: Conf. Series **915** (2017) 012007 doi:10.1088/1742-6596/915/1/012007

Solving the above equation for $\boldsymbol{F}_\infty$, we obtain

$$\boldsymbol{F}_\infty = \left(\mathbf{1} - \boldsymbol{A}_G - \tilde{\boldsymbol{B}}_{ex}^T \tilde{\boldsymbol{M}}^{-1} \tilde{\boldsymbol{B}}\right)^{-1} \boldsymbol{F}_0 . \tag{23}$$

At the origin, we use the condition

$$\boldsymbol{F}_0 = \begin{pmatrix} \mathcal{F}^{(+)}(0) \\ \mathcal{F}^{(-)}(0) \end{pmatrix} = \mathbf{1} . \tag{24}$$

For the detailed definition of the matrix in the above equations, please see Ref. [5]. The mesh points and weights of the Gauss quadrature are shown in figure 3. Also we show the scaling factor dependence of the JFM with Lagrange-mesh method and compare with the ordinary Runge-Kutta method in figure 4.

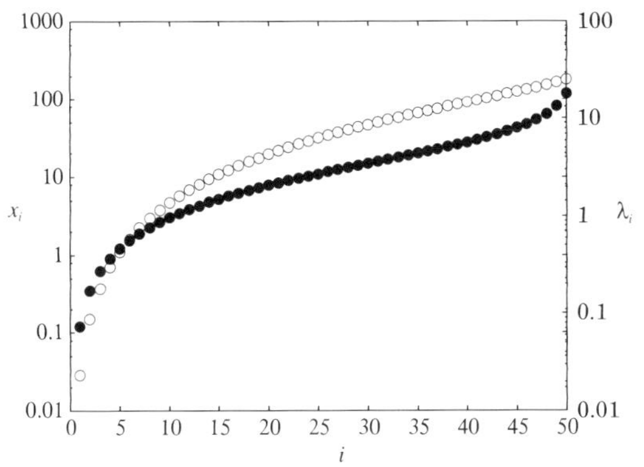

Figure 3. Zeros x_i (open circles) and weights λ_i (solid circles) of the Lagrange-Laguerre mesh.

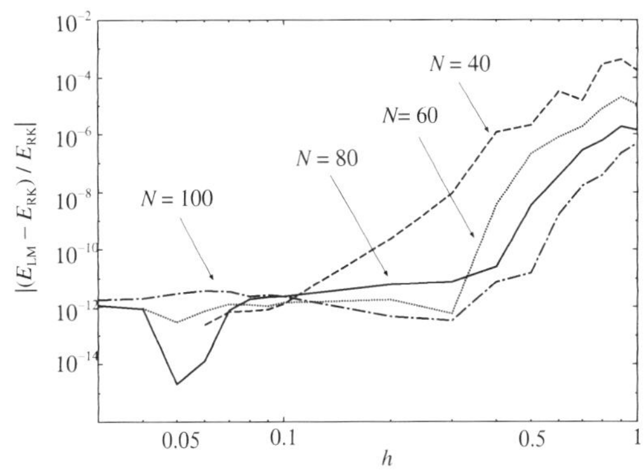

Figure 4. .Relative errors between the Lagrange-mesh ($E_{\rm LM}$) and Runge-Kutta method ($E_{\rm RK}$) in the energies.

As a numerical example of the JFM with Lagrange-mesh method, we calculate the resonant poles of the Noro-Taylor model. The results are shown in table 4, and our approach of the Lagrange-mesh method gives a reasonable agreement to those obtained by the Runge-Kutta method. The reduction of the mesh points from Runge-Kutta to Lagrange-mesh is remarkable, e.g. an order of 10,000 to at most 100. Therefore, this method can be considered as promising for solving many coupled-channels systems.

Table 4. Complex eigenvalues for the Noro-Taylor model [6]. All units for the eigenvalues are in a.u.

State	JFM-LM (this work)	Ref. [2]
1	$4.768197 - i7.10096 \times 10^{-4}$	$4.768197 - i7.10096 \times 10^{-4}$
2	$7.241200 - i7.55956 \times 10^{-1}$	$7.241200 - i7.55956 \times 10^{-1}$
3	$8.171217 - i3.25417$	$8.171216 - i3.25417$
4	$8.440530 - i6.28146$	$8.440526 - i6.28150$
5	$8.072768 - i9.57315$	$8.072642 - i9.57282$

4. Summary

We have developed the formalism of the Jost function method (JFM). Important quantities for the study of unstable nuclei, such as the partial decay widths, virtual states and non-local potentials can be discussed very accurately.

Our formalism for the partial decay width in the coupled-channel obtained using the ratio can avoid the singularity of the S-matrix at the complex eigen energy, and valid even for a very broad resonant states.

The position of the poles on the complex momentum plane gives an important information for the unstable nuclei. We showed the results for single- and coupled-channel cases for one of the typical unstable nuclei ^{10}Li. We investigated the applicability of JFM to the non-local potential in ^{10}Li case and obtained the result which agrees to the complex scaling method.

We combine the Jost function method with the Lagrange-mesh method (JFM-LM) Since the JFM-LM approach gives an accurate solution even for coupled-channel systems with 100 mesh points, it can be expected that the JFM-LM is applicable for solving the equation of the continuum discretized coupled-channel (CDCC) [19, 20]. Even though the typical number of the coupled-channel becomes more than 200 in CDCC, the total dimension of the matrix of the coupled-channel system of JFM-LM becomes $100 \times 200 = 20,000$. That is still feasible in practical calculations. Furthermore, for the study of many-body systems, the formalism of the JFM is applicable to the equations obtained with an expansion in hyper-spherical harmonics.

Acknowledgments

The author H. M would like to thank the collaborators in the study of Jost function method, K. Katō, S. Aoyama, T. Myo, and D. Baye.

References

[1] Sofianos S A and Rakityansky S A 1997 *J. PHys.* A **30** 3725; 1997 *J. PHys.* A **31** 5149
[2] Masui H, Aoyama S, Myo T and Katō K 1999 *Prog. Theor. Phys.* **102** 1119
[3] Masui H, Aoyama S, Myo T, Katō K and Ikeda K 2000 *Nucl. Phys.* **A673**, 207
[4] Masui H, Kurokawa C and Katō K 2003 *Prog. Theor. Phys.* **110** 233
[5] Masui H, Aoyama S and Baye D 2013 *Prog. Theor. Exp. Phys.* **2013** 123A02
[6] Noro T and Taylor H S 1980 *J. of Phys.* **B13** L377
[7] Moiseyev N and Peskin U, 1990 *Phys. Rev.* A **42** 255
[8] Kanada H, Kaneko T, Nagata S and Nomoto M, 1979 *Prog. Theor. Phys.* **61** 1327
[9] Thompson I J and Zhukov M V 1994 *Phys. Rev.* C **49** 1904
[10] Ho Y K 1983 *Phys. Rept.* **99** 1
[11] Aoyama S, Myo T, Katō K and Ikeda K 2006 *Prog. Theor. Phys.* **116** 1
[12] Myo T, Kikuchi Y, Masui H and Katō K 2014 *Prog. Part. Nucl. Phys.* **79** 1
[13] Katō K, Yamada T and Ikeda K 1999 *Prog. Theor. Phys.* **101** 199
[14] Baye D and Heenen P-H 1986 *J. Phys.* A **19** 2041
[15] Vincke M, Malegat L and Baye D 1993 *J. Phys.* B **26** 811
[16] Baye D and Vincke M 1999 *Phys. Rev.* E **59** 7195
[17] Baye D, Hesse M and Vincke M 2002 *Phys. Rev.* E **65** 026701
[18] Baye D 2006 *Phys. Stat. Sol.* (b) **243** 1095
[19] Kamimura M, Yahiro M, Iseri Y, Sakuragi Y, Kameyama H and Kawai M 1986 *Prog. Theor. Phys. Suppl.* **89** 1
[20] Austern N, Iseri Y, Kamimura M, Kawai M, Rawitscher G and Yahiro M 1987 *Phys. Rep.* **154** 125

IOP Conf. Series: Journal of Physics: Conf. Series **915** (2017) 012008 doi:10.1088/1742-6596/915/1/012008

Partial widths of a multi-channel resonance

Sergei Rakityansky

Dept. of Physics, University of Pretoria, Pretoria, South Africa

E-mail: `rakitsa@up.ac.za`

Abstract. A new method for finding the partial decay-widths for multi-channel resonances is suggested. It is based on using the Jost matrices. These matrices can be either calculated (when the multi-channel potential is known) or found by fitting experimental data (when they are available). In both cases the procedure for obtaining the partial widths is the same and is simple. It does not require any kind of integration or differentiation. The partial widths sum up to the total width and the corresponding branching ratios can be found as some algebraic combinations of the elements of the Jost matrices.

1. Introduction

In the quantum world, practically all scattering processes are multi-channel. This means that a collision of the quantum systems A and B may result in many different final products (channels)[1]. At certain collision energies, such a process goes via the formation of an intermediate quasi-bound state R which has a finite lifetime. This unstable state is a quantum resonance. It decays into all open channels with different probabilities,

$$A+B \longrightarrow R \longrightarrow \left\{ \begin{array}{c} A+B\ , \\ C+D\ , \\ E+F\ , \\ \text{etc}\ . \end{array} \right. \tag{1}$$

After its formation, the resonance may randomly decay in any direction, irrespective of where the original systems A and B came from, i.e in a sense it "forgets" how it was created. Mathematically, this means that the asymptotic behaviour of the corresponding wave function involves only the outgoing waves.

The half-life $T_{1/2} = (\hbar/\Gamma)\ln 2$ of the resonance R is determined by its total width Γ. If at the initial instant of time, $t = 0$, the number of generated resonances is $N(0)$, then it exponentially diminishes as

$$N(t) = N(0)\exp\left(-\frac{\Gamma}{\hbar}t\right)\ . \tag{2}$$

For each channel n, there is its own width Γ_n, which is called partial width. It is easy to see that the partial widths are fractions of the total width and must add up to Γ. Indeed, for several concurrent decay processes (1), the function $N(t)$ is a solution of the following differential equation:

$$\frac{dN}{dt} = -\frac{\Gamma_1}{\hbar}N - \frac{\Gamma_2}{\hbar}N - \cdots = -\frac{1}{\hbar}\left(\Gamma_1 + \Gamma_2 + \cdots\right)N\ , \tag{3}$$

[1] We restrict our consideration to two-body channels.

from where it follows that

$$\Gamma = \Gamma_1 + \Gamma_2 + \cdots \ . \tag{4}$$

There are many different ways of locating resonances (see, for example, Ref. [1]). The most rigorous and accurate of them are those which are based on the fact that each resonance corresponds to a pole of the S-matrix at a complex energy,

$$\mathcal{E} = \mathcal{E}^{(r)} - \frac{i}{2}\Gamma \ . \tag{5}$$

All the existing methods are able to find the resonance energy, $\mathcal{E}^{(r)}$, and the total width to a good accuracy. The disagreements among the methods show up when it comes to the partial widths.

There are two big groups of different approaches to finding the partial widths. The methods of the first group (see, for, example, Ref. [2]) are based on the ideas similar to the so-called "Fermi golden rule" which can give the amplitude of the transition from the resonant state $|R\rangle$ to a continuum state $|\psi_n\rangle$ in the channel n. Within such an approach, the corresponding partial width can be found as

$$\Gamma_n = 2\pi \left|\langle\psi_n|V|R\rangle\right|^2 \ , \tag{6}$$

where V is the interaction potential. The partial width defined as in Eq. (6), can also be obtained from the residue of the S-matrix,

$$S(E) = 1 - 2\pi i T(E) = 1 - 2\pi i \left[V + V(E-H)^{-1}V\right] \ ,$$

at the pole when $E \to \mathcal{E}$. Writing the propagator in the form of the Berggren spectral expansion [3],

$$(E-H)^{-1} = \sum_b \frac{|b\rangle\langle b|}{E-\mathcal{E}_b} + \sum_q \frac{|q\rangle\langle q|}{E-\mathcal{E}_q} + \mathcal{G}(E) \ , \tag{7}$$

which involves singular terms corresponding to bound states $|b\rangle$ and quasi-bound (resonant) states $|q\rangle$ as well as a background term $\mathcal{G}(E)$, we see that near a resonant pole, the dominant contribution to the S-matrix comes from a single term,

$$\langle\psi_n|S(E)|\psi_{n'}\rangle \underset{E\to\mathcal{E}}{\longrightarrow} -2\pi i \frac{\langle\psi_n|V|R\rangle\langle R|V|\psi_{n'}\rangle}{E-\mathcal{E}^{(r)}+\frac{i}{2}\Gamma} = \frac{-i\sqrt{\Gamma_n\Gamma_{n'}}}{E-\mathcal{E}^{(r)}+\dfrac{i}{2}\Gamma} \ , \tag{8}$$

and the absolute value of the residue coincides with $\sqrt{\Gamma_n\Gamma_{n'}}$. It was noted in many publications (see, for example, Refs. [4, 5]) that within this "golden rule" approach the partial widths obey Eq. (4) only for narrow resonances, i.e. for short-lived states they do not sum up to the total width. In Sec. 2 and 4, it is given a possible explanation of such a discrepancy.

The second group unites the methods, where the partial widths are extracted from the asymptotic behaviour of the resonant state (at large distance r between the decay fragments),

$$|R\rangle \underset{r\to\infty}{\longrightarrow} \begin{pmatrix} A_1|\psi_1\rangle \\ A_2|\psi_2\rangle \\ \vdots \end{pmatrix} \ . \tag{9}$$

In this asymptotics the amplitude A_n determines the probability of finding the system in the n-th channel and thus is related to Γ_n (see, for example, Ref. [6]).

Our approach belongs to the second group. The novelty of our method mainly consists in the way of finding the amplitudes A_n. We evaluate them using the multi-channel Jost matrices. These matrices can either be calculated from a given potential or fitted to any available experimental data. After obtaining the Jost matrices, the partial widths are calculated as simple algebraic combinations of their matrix-elements. This procedure does not involve any integration or differentiation and does not require "proper" normalization of the resonance wave function.

IOP Conf. Series: Journal of Physics: Conf. Series **915** (2017) 012008 doi:10.1088/1742-6596/915/1/012008

2. Partial widths in an experiment

Experimental equipment is always a macroscopic object. Typical distances traveled by the decay products before they are detected, are many orders of magnitude greater than the size of the quasi-bound state from which they are emitted. It is therefore possible to describe their motion towards the detectors using such notions of classical mechanics as trajectories and velocities.

Suppose at time $t = 0$, the number of generated resonances was $N(0)$. The counters registering the products of their decays are at a macroscopic distance r. Since the open channels have different thresholds, $E_1 < E_2 < E_3 \ldots$, the kinetic energies, $E_{\text{kin}}^{(n)} = E - E_n$, of the decay products are different for the channels. In the channels n and n' the products move towards the counters with the speeds v_n and $v_{n'}$. If $v_n > v_{n'}$, then a pair of simultaneous signals (for the channel-products n and n') means that the particles n were emitted from the resonance at a later time than the particles n'. The time difference is

$$\Delta t = \frac{r}{v_{n'}} - \frac{r}{v_n} = \frac{r(v_n - v_{n'})}{v_n v_{n'}} \,. \tag{10}$$

If we deal with short-lived resonances, their number (number of the emitters) during the time interval Δt can be significantly reduced. If the particles n' were emitted at $t = 0$, then the intensity of their flux was determined by the number $N(0)$ of the emitters, while for the particles n (which arrive to the detectors simultaneously with n') the number of available emitters was smaller, namely,

$$N(\Delta t) = N(0) \exp\left(-\frac{\Gamma}{\hbar}\Delta t\right) \,,$$

and thus their flux was also reduced. Therefore when trying to obtain the branching ratios $\Gamma_n/\Gamma_{n'}$ by comparing the particle-fluxes in the channels n and n', we should put the corresponding detectors at different distances, r_n and $r_{n'}$, from the emitters. These distances can be found as[2]

$$r_n = \frac{\hbar k_n}{\mu_n} t \,, \tag{11}$$

where k_n and μ_n are the wave number and the reduced mass in the channel n, while t is common time of flight (from the emitter to the detector) for all the channels. In a theoretical analysis, this requirement can be fulfilled if we consider the asymptotic behaviour of the resonance wave function (9) at the corresponding distances (11) for different channels. This is what we do in Section 4. The idea of using different asymptotic distances for different channels was taken from Ref. [6].

As was mentioned earlier, in the methods based on the "golden rule" the partial widths exactly sum up to the total width only for narrow resonances. Within these methods, the greater is Γ, the larger is the difference between the left and right hand sides of Eq. (4). The above reasoning may give a possible explanation of such a discrepancy. Indeed, the ratio $N(0)/N(\Delta t)$ exponentially grows with Γ. As we see, this error can be exactly compensated by considering the asymptotics (9) at different distances. But in the integral (6) of the "golden rule" approach there is no possibility for such a correction.

3. Multi-channel wave function

For the sake of clarity, we consider here a simplified problem, namely, the decay of a resonance generated in a two-channel system of spinless particles that interact via non-singular short-range potentials. Its generalization for an arbitrary number of channels and non-zaro spins can be done easily. The Coulomb forces can also be included without much difficulties.

[2] Of course, for fast-moving particles the relativistic kinematics should be used.

The two-channel wave function Ψ has two components, ψ_1 and ψ_2, that describe the motion in each of the channels. It is convenient to combine them in a column matrix, $\Psi = (\psi_1, \psi_2)^T$, then the Schrödinger equation, $H\Psi = E\Psi$, can be written in a matrix form. When doing the standard partial-wave decomposition of this equation, we treat each partial wave as a separate channel (even if their thresholds are the same). This means that in each of the channels the three-dimensional wave function can be written as

$$\psi_n(E, \vec{r}) = \frac{u_n(E, r)}{r} Y_{\ell_n m_n}(\hat{\vec{r}}) \ , \tag{12}$$

This gives the following system of matrix radial-equations:

$$\begin{pmatrix} \partial_r^2 + k_1^2 - \dfrac{\ell_1(\ell_1+1)}{r^2} & 0 \\ 0 & \partial_r^2 + k_2^2 - \dfrac{\ell_2(\ell_2+1)}{r^2} \end{pmatrix} \begin{pmatrix} u_1 \\ u_2 \end{pmatrix} = \begin{pmatrix} V_{11} & V_{12} \\ V_{21} & V_{22} \end{pmatrix} \begin{pmatrix} u_1 \\ u_2 \end{pmatrix} , \tag{13}$$

where the channel momentum $\hbar k_n$ (that determines the relative kinetic energy in the channel n) is given by

$$k_n = \sqrt{\frac{2\mu_n}{\hbar^2}(E - E_n)} \ , \tag{14}$$

with μ_n and E_n being the reduced mass and the threshold energy in that channel. The matrix V is the angular integral,

$$V_{nn'}(r) = \frac{2\mu_n}{\hbar^2} \int Y^*_{\ell_n m_n}(\hat{\vec{r}}) U_{nn'}(\vec{r}) Y_{\ell'_{n'} m'_{n'}}(\hat{\vec{r}}) d\hat{\vec{r}} \ ,$$

of the three-dimensional potential-matrix $U(\vec{r})$.

A system of N differential equations of the second order has $2N$ linearly independent solutions. In our case $N = 2$ and thus Eq. (13) has four independent column-solutions. It is easy to find all of them at large values of r where the (short-range) potential matrix $V(r)$ vanishes. Without the right-hand side, it becomes the Riccati-Bessel equation. Of course the choice of the independent solutions is not unique (like with any set of basis vectors). The most convenient for us is the following choice:

$$\begin{bmatrix} h^{(+)}_{\ell_1}(k_1 r) \\ 0 \end{bmatrix}_1 , \quad \begin{bmatrix} 0 \\ h^{(+)}_{\ell_2}(k_2 r) \end{bmatrix}_2 , \quad \begin{bmatrix} h^{(-)}_{\ell_1}(k_1 r) \\ 0 \end{bmatrix}_3 , \quad \begin{bmatrix} 0 \\ h^{(-)}_{\ell_2}(k_2 r) \end{bmatrix}_4 , \tag{15}$$

where $h^{(\pm)}_\ell(z)$ are the Riccati-Hankel functions. These four columns constitute a basis in the space of the solutions of Eq. (13) at large distances, i.e. any other solution is their linear combination. The first two of these columns describe the outgoing spherical waves, while the second pair represents the incoming waves. At large distances, a resonant state may only involve the outgoing waves, i.e. is a combination of the first two columns of (15). Therefore

$$\begin{pmatrix} u_1 \\ u_2 \end{pmatrix} \underset{r\to\infty}{\longrightarrow} \begin{bmatrix} A_1 h^{(+)}_{\ell_1}(k_1 r) \\ A_2 h^{(+)}_{\ell_2}(k_2 r) \end{bmatrix} \underset{r\to\infty}{\longrightarrow} \begin{bmatrix} A_1(-i)^{\ell_1+1} e^{ik_1 r} \\ A_2(-i)^{\ell_2+1} e^{ik_2 r} \end{bmatrix} , \tag{16}$$

where A_1 and A_2 are the combination coefficients. The values of them are not arbitrary. They are determined by the dynamics of the system and by a common factor that follows from the chosen normalization. As we will see, the normalization does not play any role in calculating the partial widths, while the relative strengths of the amplitudes A_n can be found using the Jost matrices.

4. Partial widths in the theory

Let us assume that the combination coefficients A_1 and A_2 in Eq. (16) have been found somehow (a way of finding them is suggested in Sec. 5). Using them, we can obtain the fluxes, $\vec{\mathcal{J}}_n = \frac{\hbar}{2\mu_n i}\left(\psi_n^*\vec{\nabla}\psi_n - \psi_n\vec{\nabla}\psi_n^*\right)$, of the decay products in each of the channels at large distances. We are only inerested in the radial component $j_n^{(r)}$ of vector $\vec{\mathcal{J}}_n = \{j_n^{(r)}, j_n^{(\theta)}, j_n^{(\varphi)}\}$, which is normal to a sphere of the radius r. Therefore, out of the three spherical components of the operator $\vec{\nabla} = \{\partial_r,\ (1/r)\partial_\theta,\ (1/r\sin\theta)\partial_\varphi\}$, we should use only the radial one. The radial flux for the function (12) with u_n given by Eq. (16), is

$$j_n^{(r)} = \frac{\hbar\mathrm{Re}(k_n)}{\mu_n r^2}|A_n|^2\exp\left[-2\mathrm{Im}(k_n)r\right] Y^*_{\ell_n m_n}(\theta,\varphi)Y_{\ell_n m_n}(\theta,\varphi)\ .$$

Then the total flux in the n-th channel through a distant sphere of the radius r is the surface integral

$$j_n = \int j_n^{(r)} r^2\sin\theta\, d\theta\, d\varphi = \frac{\hbar\mathrm{Re}(k_n)}{\mu_n}|A_n|^2\exp\left[-2\mathrm{Im}(k_n)r\right]\ . \tag{17}$$

At this point we should recall the discussion given in Section 2 about the choice for each channel a distances r_n which is covered by the decay products during the same time t in all the channels. For a resonance the momentum $\hbar k_n$ is complex. The speed of the products is determined by its real part, and thus

$$r_n = \frac{\hbar\mathrm{Re}(k_n)}{\mu_n}t\ . \tag{18}$$

At this distance the flux (17) is

$$j_n = \frac{\hbar\mathrm{Re}(k_n)}{\mu_n}|A_n|^2\exp\left[-2\mathrm{Im}(k_n)\frac{\hbar\mathrm{Re}(k_n)}{\mu_n}t\right] = \frac{\hbar\mathrm{Re}(k_n)}{\mu_n}|A_n|^2\exp\left(\frac{\Gamma}{\hbar}t\right)\ , \tag{19}$$

where we used the fact that the total width Γ is expressed via the channel momentum as (this can be seen if the energy (5) is substituted in Eq. (14) instead of E)

$$\Gamma = -\frac{2\hbar^2\mathrm{Re}(k_n)\mathrm{Im}(k_n)}{\mu_n} \qquad \text{for any} \quad n\ . \tag{20}$$

For all the channels the time-dependent factor in Eq. (19) is the same. Therefore in the ratios $j_n/j_{n'}$ it is always canceled out. Apparently, such a ratio is equal to the ratio of the corresponding decay rates, i.e. it is the branching ratio,

$$\frac{\Gamma_n}{\Gamma_{n'}} = \frac{j_n}{j_{n'}} = \frac{\mu_{n'}\mathrm{Re}(k_n)|A_n|^2}{\mu_n\mathrm{Re}(k_{n'})|A_{n'}|^2}\ . \tag{21}$$

The fluxes given by Eq. (19), were originated from the emitter at the same instant of time in all the channels. Therefore the channel flux j_n for a specific n is a fraction f_n of the total flux originated at that time,

$$f_n = \frac{j_n}{j_1 + j_2 + \cdots}\ , \tag{22}$$

and the channel width Γ_n is the same fraction of the total width,

$$\Gamma_n = f_n\Gamma = \frac{\mathrm{Re}(k_n)|A_n|^2\Gamma}{\sum\limits_{n'=1}^{N}\frac{\mu_n}{\mu_{n'}}\mathrm{Re}(k_{n'})|A_{n'}|^2}\ , \tag{23}$$

where N is the number of open channels. In the wave function (16), all the amplitudes A_n have a common normalization factor, which is canceled out in Eq. (23). We therefore avoid the problem of "proper" normalization of the resonant wave function. What remains is to find a way of obtaining the channel amplitudes A_n. This is done in Section 6.

At this point a question arises about Eq. (8), from which it follows that $\Gamma_n = |\mathrm{Res}(S_{nn}, \mathcal{E})|$. This cannot be completely wrong and thus the residue of the diagonal element of the S-matrix at the resonance pole must somehow be related the the corresponding channel amplitude A_n. And indeed such a relation can be found (see, for example, Ref. [7]),

$$\mathrm{Res}(S_{nn}, \mathcal{E}) = \frac{i\hbar^2 k_n}{\mu_n}(-1)^{\ell_n+1}\,|A_n|^2 \ , \tag{24}$$

from which we obtain

$$\left|\frac{\mathrm{Res}\, S_{nn}}{\mathrm{Res}\, S_{n'n'}}\right| = \frac{\mu_{n'}|k_n|\,|A_n|^2}{\mu_n|k_{n'}|\,|A_{n'}|^2} \ . \tag{25}$$

Comparing this equation with Eq. (21), we see that the above ratio of the residues differs from the ratio of the corresponding partial widths by a kinematical factor,

$$\frac{\Gamma_n}{\Gamma_{n'}} = \left|\frac{\mathrm{Res}\, S_{nn}}{\mathrm{Res}\, S_{n'n'}} \cdot \frac{k_{n'}\mathrm{Re}(k_n)}{k_n\mathrm{Re}(k_{n'})}\right| \ . \tag{26}$$

This kinematical factor, $k_{n'}\mathrm{Re}(k_n)/[k_n\mathrm{Re}(k_{n'})]$, stems from the differences in the time of flight in different channels (discussed in Sec. 2). It becomes noticeable for wide resonances, i.e. when the difference between $|k_n|$ and $\mathrm{Re}(k_n)$ is significant.

Therefore the numerator of Eq. (8) cannot be merely written as $\sqrt{\Gamma_n\Gamma_{n'}}$. It should involve some kinematical factors. These factors may be more complicated than those given in Eq. (26) because in the ratio some of them may cancel. Such kinematical factors must appear also in the "golden rule" formula (6) as a result of a proper normalization of the states $|R\rangle$ and $|\psi_n\rangle$. As we see, the factors become insignificant for narrow resonances. This fact perhaps explains why the "golden rule" produces sufficiently accurate results for them and becomes inaccurate when Γ increases [4, 5].

5. Jost matrices

We assumed that the potential V is not singular. This means that all of its matrix elements are regular or (in the worst case) less singular than $\sim r^{-2}$ at $r = 0$. Therefore the most severe singularity of Eq. (13) is due to the centrifugal terms. However if we multiply the equation by r^2, none of its coefficients remain singular at $r = 0$. In the theory of differential equations, such points are called the regular singular-points (see, for example, Ref. [8]).

A system of N second-order differentia equations has $2N$ linearly independent solutions. At a regular singular-point, half of these solutions are regular while the other half may diverge at such a point. Each solution of Eq. (13) is a column. Therefore there are four linearly independent columns that solve this equation,

$$\begin{pmatrix}\phi_1\\ \phi_2\end{pmatrix}_1 , \quad \begin{pmatrix}\phi_1\\ \phi_2\end{pmatrix}_2 , \quad \begin{pmatrix}\varphi_1\\ \varphi_2\end{pmatrix}_1 , \quad \begin{pmatrix}\varphi_1\\ \varphi_2\end{pmatrix}_2 ,$$

where the symbols ϕ and φ are used for the regular and irregular solutions, respectively. These columns can be combined in two square matrices,

$$\phi = \begin{pmatrix}\phi_{11} & \phi_{12}\\ \phi_{21} & \phi_{22}\end{pmatrix} , \quad \varphi = \begin{pmatrix}\varphi_{11} & \varphi_{12}\\ \varphi_{21} & \varphi_{22}\end{pmatrix} ,$$

where the second subscript corresponds to the solution number. Since we are only interested in a solution that is relevant to physics, we may forget about the matrix φ and only consider the matrix ϕ whose columns are regular at the point $r = 0$. These columns form a basis in the space of regular solutions. This means that the physical solution u is their linear combination,

$$\begin{pmatrix} u_1 \\ u_2 \end{pmatrix} = \begin{pmatrix} \phi_1 \\ \phi_2 \end{pmatrix}_1 C_1 + \begin{pmatrix} \phi_1 \\ \phi_2 \end{pmatrix}_2 C_2 = \begin{pmatrix} \phi_{11}C_1 + \phi_{12}C_2 \\ \phi_{21}C_1 + \phi_{22}C_2 \end{pmatrix} = \begin{pmatrix} \phi_{11} & \phi_{12} \\ \phi_{21} & \phi_{22} \end{pmatrix} \begin{pmatrix} C_1 \\ C_2 \end{pmatrix} , \tag{27}$$

where the combination coefficients C_1 and C_2 should be chosen in such a way that the function u has a correct asymptotic behaviour (when $r \to \infty$).

At large distances, where the right-hand side of Eq. (13) vanishes, it still has four linearly independent solutions, namely, the columns (15). When $r \to \infty$, there is no requirement of regularity. The four columns (15) constitute a full basis there, and any other solutions is a linear combination of all of them. In particular, our regular basis basis ϕ at large distances is also a combination of them,

$$\begin{pmatrix} \phi_{1n} \\ \phi_{2n} \end{pmatrix} \underset{r\to\infty}{\longrightarrow} \begin{bmatrix} h_{\ell_1}^{(-)} \\ 0 \end{bmatrix} f_{1n}^{(\mathrm{in})} + \begin{bmatrix} 0 \\ h_{\ell_2}^{(-)} \end{bmatrix} f_{2n}^{(\mathrm{in})} + \begin{bmatrix} h_{\ell_1}^{(+)} \\ 0 \end{bmatrix} f_{1n}^{(\mathrm{out})} + \begin{bmatrix} 0 \\ h_{\ell_2}^{(+)} \end{bmatrix} f_{2n}^{(\mathrm{out})} , \qquad n = 1, 2 ,$$

where the combination coefficients $f_{nn'}^{(\mathrm{in/out})}$ have two subscripts: the first one shows in which channel the column (15) has a non-zero element, and the second subscript indicates which of the columns of the regular matrix is expanded. Combining the regular columns in the square matrix, we have

$$\begin{pmatrix} \phi_{11} & \phi_{12} \\ \phi_{21} & \phi_{22} \end{pmatrix} \underset{r\to\infty}{\longrightarrow} \begin{bmatrix} h_{\ell_1}^{(-)} f_{11}^{(\mathrm{in})} & h_{\ell_1}^{(-)} f_{12}^{(\mathrm{in})} \\ h_{\ell_2}^{(-)} f_{21}^{(\mathrm{in})} & h_{\ell_2}^{(-)} f_{22}^{(\mathrm{in})} \end{bmatrix} + \begin{bmatrix} h_{\ell_1}^{(+)} f_{11}^{(\mathrm{out})} & h_{\ell_1}^{(+)} f_{12}^{(\mathrm{out})} \\ h_{\ell_2}^{(+)} f_{21}^{(\mathrm{out})} & h_{\ell_2}^{(+)} f_{22}^{(\mathrm{out})} \end{bmatrix}$$

$$= \begin{bmatrix} h_{\ell_1}^{(-)} & 0 \\ 0 & h_{\ell_2}^{(-)} \end{bmatrix} \begin{bmatrix} f_{11}^{(\mathrm{in})} & f_{12}^{(\mathrm{in})} \\ f_{21}^{(\mathrm{in})} & f_{22}^{(\mathrm{in})} \end{bmatrix} + \begin{bmatrix} h_{\ell_1}^{(+)} & 0 \\ 0 & h_{\ell_2}^{(+)} \end{bmatrix} \begin{bmatrix} f_{11}^{(\mathrm{out})} & f_{12}^{(\mathrm{out})} \\ f_{21}^{(\mathrm{out})} & f_{22}^{(\mathrm{out})} \end{bmatrix} .$$

It is not difficult to see that for an arbitrary N, we have the following asymptotic behaviour of the matrix of regular solutions,

$$\phi(E, r) \underset{r\to\infty}{\longrightarrow} W^{(\mathrm{in})}(E, r) f^{(\mathrm{in})}(E) + W^{(\mathrm{out})}(E, r) f^{(\mathrm{out})}(E) , \tag{28}$$

where $W^{(\mathrm{in/out})}$ are the diagonal matrices

$$W^{(\mathrm{in/out})}(E, r) = \mathrm{diag}\left\{ h_{\ell_1}^{(\mp)}(k_1 r), h_{\ell_2}^{(\mp)}(k_2 r), \ldots, h_{\ell_N}^{(\mp)}(k_N r) \right\} , \tag{29}$$

describing the incoming and outgoing multi-channel spherical waves, and the energy-dependent square matrices $f^{(\mathrm{in/out})}$ composed of the combination coefficients, are the multi-channel Jost matrices.

6. Asymptotic behaviour of a resonant solution

Consider the physical solution (27) at large distances,

$$u = \phi C \underset{r\to\infty}{\longrightarrow} W^{(\mathrm{in})} f^{(\mathrm{in})} C + W^{(\mathrm{out})} f^{(\mathrm{out})} C , \tag{30}$$

where C is the column-matrix, $C = (C_1, C_2)^T$. By definition, a spectral point (corresponding to either bound or resonant state) is a value of the energy, $E = \mathcal{E}$, for which the wave function has only the outgoing waves in its asymptotic behaviour. From Eq. (30) it follows that this can be achieved if

$$\begin{bmatrix} f_{11}^{(\text{in})} & f_{12}^{(\text{in})} \\ f_{21}^{(\text{in})} & f_{22}^{(\text{in})} \end{bmatrix} \begin{pmatrix} C_1 \\ C_2 \end{pmatrix} = 0 \,. \tag{31}$$

It is a homogeneous linear system of equations for unknown C_1 and C_2. It has a non-trivial solution if and only if

$$\det f^{(\text{in})}(\mathcal{E}) = 0 \,. \tag{32}$$

Complex roots of this equation correspond to the resonance energies (5). After finding such a root, we can solve the homogeneous system of linear equations (31) for the unknown coefficients C_1 and C_2. In the two-channel case this is trivial: C_1 is arbitrary and $C_2 = -C_1 f_{11}^{(\text{in})}/f_{12}^{(\text{in})}$, where C_1 could (in principle) be fixed from the normalization condition. However, as we will see, any common normalization factor is canceled out in the ratio (23).

Now, according to Eq. (30), the resonance-state wave function asymptotically behaves as

$$u(\mathcal{E}, r) \underset{r\to\infty}{\longrightarrow} W^{(\text{out})}(\mathcal{E}, r) f^{(\text{out})}(\mathcal{E}) C \tag{33}$$

$$= \begin{bmatrix} h_{\ell_1}^{(+)}(k_1 r) & 0 \\ 0 & h_{\ell_2}^{(+)}(k_2 r) \end{bmatrix} \begin{bmatrix} f_{11}^{(\text{out})}(\mathcal{E}) & f_{12}^{(\text{out})}(\mathcal{E}) \\ f_{21}^{(\text{out})}(\mathcal{E}) & f_{22}^{(\text{out})}(\mathcal{E}) \end{bmatrix} \begin{bmatrix} C_1 \\ -\dfrac{f_{11}^{(\text{in})}(\mathcal{E})}{f_{12}^{(\text{in})}(\mathcal{E})} C_1 \end{bmatrix} = \begin{bmatrix} A_1 h_{\ell_1}^{(+)}(k_1 r) \\ A_2 h_{\ell_2}^{(+)}(k_2 r) \end{bmatrix} ,$$

where A_1 and A_2,

$$A_1 = C_1 \left[f_{11}^{(\text{out})} - \frac{f_{11}^{(\text{in})} f_{12}^{(\text{out})}}{f_{12}^{(\text{in})}} \right]_{E=\mathcal{E}} , \qquad A_2 = C_1 \left[f_{21}^{(\text{out})} - \frac{f_{11}^{(\text{in})} f_{22}^{(\text{out})}}{f_{12}^{(\text{in})}} \right]_{E=\mathcal{E}} , \tag{34}$$

are the same asymptotic amplitudes of the channels that appear in Eqs. (16) and (23). As is seen, the common factor C_1 can be canceled in Eq. (23). We therefore can put it to unity, $C_1 = 1$.

7. Finding the Jost matrices

The partial widths Γ_n can be found by substituting the channel amplitudes (34) into Eq. (23). Such a calculation requires the knowledge of the Jost matrices $f^{(\text{in})}(\mathcal{E})$ and $f^{(\text{out})}(\mathcal{E})$ at the resonance energy (5). We can obtain these matrices either using the interaction potential (when it is known) or by fitting available experimental data.

7.1. Jost matrices from the potential

If the matrix $V(r)$ of the potential is known, the Jost matrices $f^{(\text{in})}(E)$ and $f^{(\text{out})}(E)$ can be calculated for any complex energy E by a numerical integration of the following system of differential equations (the derivation of these equations and all the details concerning their solutions can be found in Refs. [9–12]):

$$\partial_r F^{(\text{in})} = -\frac{1}{2i} K^{-1} W^{(\text{out})} V \left[W^{(\text{in})} F^{(\text{in})} + W^{(\text{out})} F^{(\text{out})} \right] , \tag{35}$$

$$\partial_r F^{(\text{out})} = \frac{1}{2i} K^{-1} W^{(\text{in})} V \left[W^{(\text{in})} F^{(\text{in})} + W^{(\text{out})} F^{(\text{out})} \right] , \tag{36}$$

where the diagonal matrix $K = \mathrm{diag}\{k_1, k_2, \ldots, k_N\}$, consists of the channel momenta, and the unknown matrices $F^{(\mathrm{in/out})}(E,r)$ asymptotically tend to the Jost matrices,

$$\lim_{r\to\infty} F^{(\mathrm{in/out})}(E,r) = f^{(\mathrm{in/out})}(E) \ . \tag{37}$$

Starting from the boundary values of these matrices at $r = 0$,

$$F^{(\mathrm{in/out})}_{nn'}(E,0) = \delta_{nn'} \ , \tag{38}$$

we can proceed (by numerical integration) to such a distance where the potential vanishes and thus the right-hand sides of Eqs. (35, 36) become zero. Since at that distance the derivatives $\partial_r F^{(\mathrm{in/out})}$ are zero, the limits (37) are reached.

There is a technical difficulty in solving Eqs. (35, 36) for complex energies. Their solutions do not converge to the limits (37) along the real radius r. However this difficulty can be easily overcome if the integration is done along a ray rotated in the complex plane of the coordinate, $r = |r|\exp(i\theta)$, with a large enough angle θ (the details can be found in Refs. [9–12]).

7.2. Jost matrices from the data

The resonance energy, total width, and partial widths can be extracted from any available experimental data by fitting these data with the help of the properly parametrized Jost matrices. In Ref. [13] it was shown that the Jost matrices have the following analytic structure:

$$f^{(\mathrm{in})}_{mn}(E) = \frac{k_n^{\ell_n+1}}{k_m^{\ell_m+1}} a_{mn}(E) - i k_m^{\ell_m} k_n^{\ell_n+1} b_{mn}(E) \ , \tag{39}$$

$$f^{(\mathrm{out})}_{mn}(E) = \frac{k_n^{\ell_n+1}}{k_m^{\ell_m+1}} a_{mn}(E) + i k_m^{\ell_m} k_n^{\ell_n+1} b_{mn}(E) \ , \tag{40}$$

where the energy-dependent matrices $a(E)$ and $b(E)$ are the same for both $f^{(\mathrm{in})}$ and $f^{(\mathrm{out})}$. In the same Ref. [13] it was established that the matrices $a(E)$ and $b(E)$ are single-valued analytic functions of the energy, defined on a single one-layer energy plane, i.e. all the complications stemming from the branch points are isolated in Eqs. (39, 40) via the explicit factors depending on the channel momenta.

Since the matrices $a(E)$ and $b(E)$ are analytic functions of the variable E, they can be expanded in the Taylor series,

$$a(E) = \sum_{n=0}^{\infty} (E-E_0)^n \alpha_n(E_0) \ , \qquad b(E) = \sum_{n=0}^{\infty} (E-E_0)^n \beta_n(E_0) \ , \tag{41}$$

near an arbitrary point E_0 within the domain of their analyticity. Here the expansion coefficients α_n and β_n are matrices of the same dimension as a and b. They depend on the choice of the point E_0.

We can approximate the matrices $a(E)$ and $b(E)$ by a finite number of terms in the expansions (41),

$$a(E) \approx \sum_{n=0}^{M} (E-E_0)^n \alpha_n(E_0) \ , \qquad b(E) \approx \sum_{n=0}^{M} (E-E_0)^n \beta_n(E_0) \ , \tag{42}$$

and consider the elements of the matrices α_n and β_n as the fitting parameters. The expansion central point E_0 can be chosen on the real axis somewhere in the middle of the available

IOP Conf. Series: Journal of Physics: Conf. Series **915** (2017) 012008 doi:10.1088/1742-6596/915/1/012008

experimental data (cross sections, analyzing power, etc.). The data can be fitted using the S-matrix

$$S(E) = f^{(\mathrm{out})}(E)\left[f^{(\mathrm{in})}(E)\right]^{-1}, \tag{43}$$

where the Jost matrices depend of the parameters α_n and β_n. After finding the optimal values of these parameters, we can use the same Jost matrices at complex energies within a circle around E_0. In this way the resonance energy (5) can be located and the channel amplitudes (34) for Eq. (23) can be found.

8. Generalization

In order to make all the derivations clear, we simplified the problem, namely,

- The system under consideration had only two channels ($N = 2$);
- The spins of the decay fragments were zero;
- The interaction potentials were of a short-range type.

The first of these limitations can be dismissed most easily. For an arbitrary N, the dimensions of all the matrices are increased accordingly. When solving Eq. (31) for the unknown coefficients $C_1, C_2, \ldots, C_N$, we can still put $C_1 = 1$ and find all the other of them via the matrix elements of $f^{(\mathrm{in})}(\mathcal{E})$. And finally, Eqs. (34) are to be replaced with more complicated (but still simple) expressions for $A_1, A_2, \ldots, A_N$.

Let the decay fragments have non-zero spins. The total angular momentum J as well as its third component M are conserving. As a result only few partial waves are mixed to each other. If $\vec{s}$ is the total spin of the interacting particles, then $\vec{J} = \vec{\ell} + \vec{s}$, and the maximal number of coupled partial waves is determined by the triangle condition $|J - s| \leqslant \ell \leqslant |J + s|$. There is also the parity conservation law, which allows mixing of either only even or only odd angular momenta ℓ.

When generalizing the partial-wave expansion (12), we should use the spin-angular functions (instead of the ordinary spherical harmonics),

$$\mathcal{Y}_{\ell s}^{JM}(\hat{\vec{r}}) = \sum_{m\mu} C_{\ell m s\mu}^{JM} Y_{\ell m}(\hat{\vec{r}})\chi_{s\mu}, \tag{44}$$

where $\chi_{s\mu}$ is the spin function depending on the spin s and its third component μ. For each open channel on the right-hand side of Eq. (1) there is its own set of possible combinations of the pair (ℓs). Each of these combinations should be considered as a separate channel. For example, if the first two channels are open, namely, $A+B$ and $C+D$, with two and three possible combinations of (ℓs), respectively,

$$\begin{array}{llcl} n=1 & A+B & : & (\ell_1 s_1)_1\,,\quad (\ell_2 s_2)_1\,, \\ n=2 & C+D & : & (\ell_1 s_1)_2\,,\quad (\ell_2 s_2)_2\,,\quad (\ell_3 s_3)_2\,, \end{array}$$

then we have to solve an effective five-channel problem. Apparently, for these effective channels all the formulae given in the previous sections remain the same. After obtaining all five partial widths, we should group them as follows:

$$\begin{array}{llcl} n=1 & A+B & : & \Gamma_1 = \Gamma_{(\ell_1 s_1)_1} + \Gamma_{(\ell_2 s_2)_1}\,, \\ n=2 & C+D & : & \Gamma_2 = \Gamma_{(\ell_1 s_1)_2} + \Gamma_{(\ell_2 s_2)_2} + \Gamma_{(\ell_3 s_3)_2}\,. \end{array}$$

The last of our simplifications was the absence of the long-range Coulombic tails of the matrix

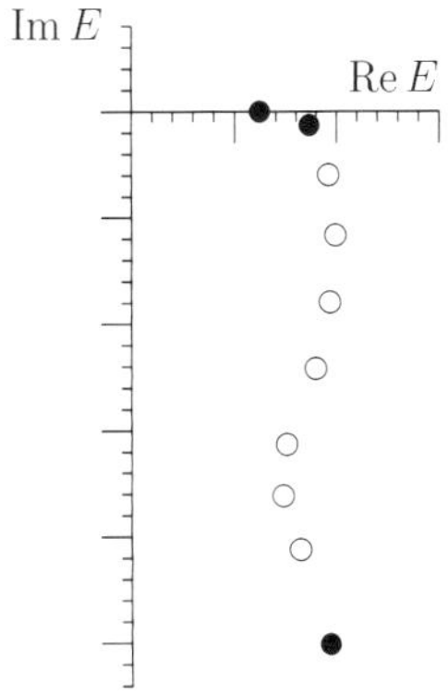

Figure 1. Resonance poles generated by the two-channel model (46, 47). For the poles shown by the filled circles, the resonance parameters are given in Table 1.

Table 1. Parameters of the three resonances shown as the filled circles in Fig. 1.

$\mathcal{E}^{(r)} - \frac{i}{2}\Gamma$	Γ_1	Γ_2	$\frac{\Gamma_1}{\Gamma_2}$	method
	0.037827	0.00049485	76.442	residues
$6.2000 - \frac{i}{2}0.038514$	0.038017	0.00049733	76.442	Masui et al.
	0.038017	0.00049733	76.442	present work
	0.18549	1.5763	0.11768	residues
$8.6519 - \frac{i}{2}1.2992$	0.13679	1.1624	0.11768	Masui et al.
	0.13685	1.1623	0.11773	present work
	58.827	2.4282	24.226	residues
$9.6786 - \frac{i}{2}50.155$	48.166	1.9882	24.226	Masui et al.
	48.214	1.9403	24.848	present work

elements of V. If the decay products have non-zero charges, we should replace the Riccati-Hankel functions $h_\ell^{(\pm)}(kr)$ with the corresponding Coulomb functions,

$$H_\ell^{(\pm)}(\eta, kr) = F_\ell(\eta, kr) \mp iG_\ell(\eta, kr) \underset{r\to\infty}{\longrightarrow} (\mp i)^{\ell+1} \exp\{\pm i\,[kr - \eta\ln(2kr) + \sigma_\ell]\} \ . \tag{45}$$

It is easy to check that with such a replacement all the formulae for calculating the partial widths remain the same. As far as the differential equations (35,36) and the analytic structure (39,40) of the Jost matrices are concerned, their generalizations for the charged particles can be found in Ref. [12]. In essence, with these generalizations the procedures for finding the Jost matrices either from a potential or by fitting the data remain the same.

9. Numerical example

In order to test the formulae derived in this paper, let us consider a simple two-channel model with the potential

$$U(r) = \begin{pmatrix} -1.0 & -7.5 \\ -7.5 & 7.5 \end{pmatrix} r^2 e^{-r} \ , \tag{46}$$

IOP Conf. Series: Journal of Physics: Conf. Series **915** (2017) 012008 doi:10.1088/1742-6596/915/1/012008

given in the units such that $\hbar c = 1$. This is the same potential that is used in the well-known Noro-Taylor model [14]. However, the original model is not sufficiently sensitive to the kinematic factor in Eq. (26). The reason is that the mass is the same in both channels, $\mu_1 = \mu_2 = 1$, and their thresholds are very close to each other, $E_1 = 0$, $E_2 = 0.1$. As a result the time of flight is practically the same in both channels. To avoid this, let us modify the model by using different masses and a larger difference between the thresholds,

$$\mu_1 = 1\ , \quad \mu_2 = 1.1\ , \quad E_1 = 0\ , \quad E_2 = 2\ . \tag{47}$$

Some resonance poles found within this modified Noro-Taylor model are shown in Fig. 1. For the two narrow and one wide resonance (filled circles in Fig. 1), the partial widths were calculated using four different methods. In the first method it was assumed that Eq. (8) is correct and thus $\Gamma_n = |\mathrm{Res}(S_{nn}, \mathcal{E})|$. The second method was suggested by Masui et al. in Ref. [15], where the authors avoided the problem of normalization in Eq. (8) by taking the ratio of the widths as $\Gamma_1/\Gamma_2 = \mathrm{Res}\, S_{11}/\mathrm{Res}\, S_{22}$ and using the condition $\Gamma_1 + \Gamma_2 = \Gamma$. They however did not include the kinematical factor that is present in Eq. (26). Within the third method, the partial widths were calculated using Eq. (26) with this factor included. And lastly, the fourth method was based on Eq. (23) where the asymptotic amplitudes A_n we calculated as is described in Sec. 7.1.

The results of the calculations are given in Table 1. It turned out that the third and the fourth methods produced identical numbers. For each resonance, they therefore are represented in the table by a common line denoted as "present work".

10. Summary and conclusion

It is shown how the partial widths of a multi-channel resonance can be found using the Jost matrices that determine the asymptotic behaviour of the resonance wave function. After the Jost matrices are obtained, the suggested method does not require any additional integration or differentiation. A normalization of the resonance wave function is not needed. The partial widths are calculated as certain fractions of the total width and thus they exactly sum up to it. These fractions are obtained as the corresponding ratios of the channel-fluxes to the total flux of the decay products. When calculating such ratios, the channel fluxes emitted at the same instant of time are considered.

The proposed method can be used for the analysis of various non-relativistic quantum resonance phenomena in nuclear, atomic, and molecular physics. Depending on the type of the task, the Jost matrices needed for finding the partial widths, can either be calculated from a given potential, or found approximately by fitting available experimental data.

References

[1] Kukulin V I, Krasnopolsky V M and Horaček J 1988 *"Theory of Resonances"*, Dortrecht: Kluwer
[2] Rafael de la Madrid 2015 *Nucl.Phys.* A **940** 297-310
[3] Berggren T 1968 *Nucl.Phys.* A **109** 265-287
[4] McCurdy C W and Rescigno T N 1979 *Phys. Rev.* A **20** 2346-2351
[5] Rescigno T N and McCurdy C W 1986 *Phys. Rev.* A **34** 1882-1887
[6] Goldzak T, Gilary I and Moiseyev N 2010 *Phys. Rev.* A **82** 052105
[7] Baz A I, Zeldovich Ya B and Perelomov A M 1969 *"Scattering, reactions and decay in nonrelativistic quantum mechanics"*, Israel Program for Scientific Translations: Jerusalem
[8] Whittaker E T, Watson G N 1950 *"A course of modern analysis"* Cambridge University Press: Cambridge
[9] Rakityansky A S and Sofianos S A 1998 *J. Phys.* A **31** 5149-5175
[10] Rakityansky S A and Sofianos S A 1999 *Few-Body Syst. Suppl.* **10** 93-96
[11] Rakityansky S A and ElanderN 2006 *Int. J. Quantum Chem.* **106** 1105-1129
[12] Rakityansky S A and Elander N 2013 *Journal of Mathematical Physics* **54** 122112
[13] Rakityansky S A and Elander N 2011 *J. Phys.* A **44** 115303
[14] Noro T and Taylor H S 1980 *J. Phys.* B **13** L377
[15] Masui H, Aoyama S, Myo T and Kato K 1999 *Progress of Theoretical Physics* **102** 1119-1131

IOP Conf. Series: Journal of Physics: Conf. Series **915** (2017) 012009 doi:10.1088/1742-6596/915/1/012009

Energy dependence of the continuum-continuum couplings in the breakup of ^{6}Li on ^{144}Sm target

B Mukeru

Department of Physics, University of South Africa, P.O. Box 392, Pretoria 0003, South Africa

E-mail: mukerb1@unisa.ac.za

Abstract. The energy dependence of the continuum-continuum couplings is studied for the ^{6}Li + ^{144}Sm breakup reaction. We show that for energies well below the Coulomb barrier ($\gamma = E_{\rm cm}/V_B \leq 0.7$), continuum-continuum couplings enhance the breakup cross sections. However, for $\gamma \geq 0.8$, the breakup cross sections are suppressed, owing to these couplings. It is shown that the suppression factor increases with the incident energy.

1. Introduction

In the recent past, dynamic effects of the continuum-continuum couplings (ccc) in breakup reactions induced by loosely bound nuclei have been intensively investigated [1, 2, 3, 4, 5, 6, 7, 8, 9]. It is generally agreed that, once such couplings are included in the coupling matrix element, they result in a substantial suppression of the breakup cross section at energies above the Coulomb barrier. Not only has the attention being focused on energies above the barrier, but also most of the results in the literature are based on halo nuclei, like ^{8}B, ^{11}Be, ^{15}C etc. Not much has been done in this direction as far as cluster nuclei like ^{6,7}Li are concerned. However, analysing the ^{7}Li + ^{144}Sm reaction in Ref. [10], it was shown that for energies below and around the barrier, the breakup cross sections are rather enhanced, owing to the ccc. It is unclear whether such conclusion could be extended to the ^{6}Li nucleus, given on one hand, their different ground state energies upon which the breakup cross section are strongly dependent. Moreover, in any case, the energy dependence of the ccc effects is yet to be fully established. In order to understand clearly the importance of the ccc in the breakup process, it is interesting to investigate their energy dependence. A detailed analysis of the ccc effects in breakup reactions could shed more light on similarities and differences in the breakup of halo and loosely bound cluster nuclei.

In this paper, we study the energy dependence of the ccc in the ^{6}Li + ^{144}Sm breakup reaction. To this end, we consider incident energy below, around and above the Coulomb barrier ($0.6 \leq E_{\rm cm}/V_B \leq 2$, where $V_B = 23.8\,$MeV is the barrier height [11]). The different breakup cross sections are obtained by means of the CDCC (Continuum Discretized Coupled-Channels) method [12], after the resulting coupled differential equations are numerically solved using FRESCO [13]. This reaction has been intensively investigated, especially for fusion cross, for example [14]. The paper is organized as follows: in Section 2, we present and discuss the results, while the conclusions are reported in Section 3.

2. Results and Discussion

The ^{6}Li prompt breakup is well modeled as $^6\text{Li} \to \alpha$+d. The parameters of the $V_{\alpha d}$ potential used to generate the ground state and bin wave functions are taken from Ref. [15]. The $V_{\alpha t}$ and V_{dt} potential parameters are the same as those used in Ref. [16], together with the different CDCC breakup model parameters. In what follows, σ_{ccc} denotes the breakup cross section in the presence of ccc and σ_{nccc} the breakup cross section in the absence of ccc. In Fig. 1, we summarize

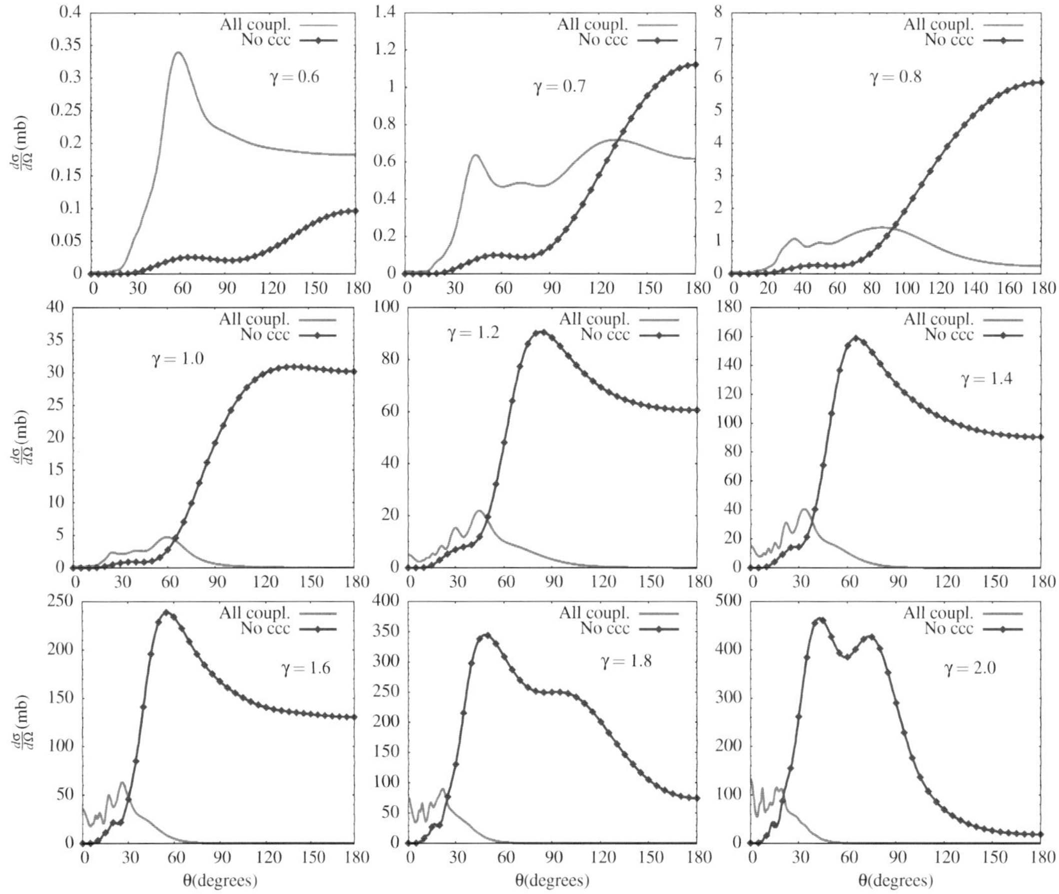

Figure 1. Differential breakup cross sections for energies below and above the Coulomb barrier.

the differential angular distributions breakup cross sections for energies below, around and above the barrier ($0.6 \leq \gamma \leq 2$, where $\gamma = E_{\text{cm}}/V_B$). Looking at this figure, one observes that for $\gamma = 0.6$, the breakup cross section σ_{ccc} is largely enhanced, owing to the ccc, since it is clear that $\sigma_{\text{ccc}} \gg \sigma_{\text{nccc}}$ for the whole angular range displayed. For $\gamma = 0.7$, we observe that the cross section is still enhanced at $\theta \leq 140°$, while beyond this angle, σ_{ccc} is suppressed due to these couplings ($\sigma_{\text{ccc}} < \sigma_{\text{nccc}}$). However, for $\gamma \geq 0.8$, we observe that at forward angles, the breakup cross sections σ_{ccc} are relatively enhanced. At backward angles, on the other hand, we notice a substantial suppression of these cross sections, because of $\sigma_{\text{ccc}} \ll \sigma_{\text{nccc}}$. One also notices that the large extension of the σ_{nccc} breakup cross sections towards larger angles is continuously washed away as the incident energy increases.

In Fig. 2, we plot the breakup cross sections σ_{ccc} and σ_{nccc} as functions γ, and show that the suppression of σ_{ccc} due to the ccc above the barrier increases with γ. We performed a numerical

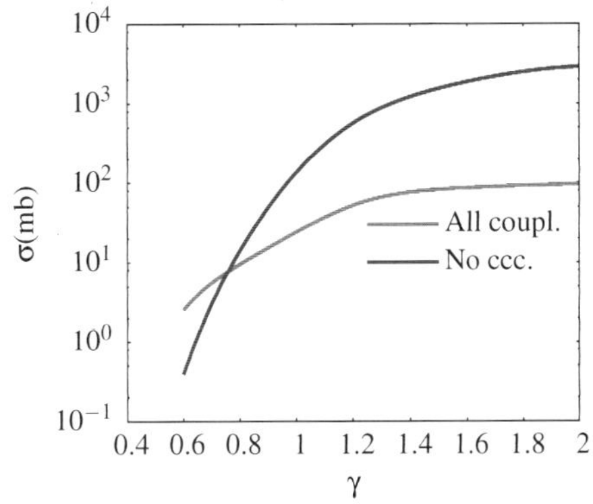

Figure 2. Integrated $\sigma_{\rm ccc}$ and $\sigma_{\rm nccc}$ breakup cross sections as functions of γ.

integration of the differential breakup cross sections and show that for $\gamma = 0.6$ the integrated breakup cross sections are $\sigma_{\rm ccc} = 2.542\,{\rm mb}$ and $\sigma_{\rm nccc} = 0.389\,{\rm mb}$, giving an enhancement factor of 0.85. For $\gamma = 0.7$, $\sigma_{\rm ccc} = 6.743\,{\rm mb}$ and $\sigma_{\rm nccc} = 4.033\,{\rm mb}$, corresponding to an enhancement factor of 0.40. However, for $\gamma = 0.8$ for example, $\sigma_{\rm ccc} = 11.50\,{\rm mb}$ while $\sigma_{\rm nccc} = 24.35\,{\rm mb}$, resulting in a suppression factor of 0.53.

3. Conclusions

In this paper, we have studied the energy dependence of the ccc in the ^{6}Li + ^{144}Sm breakup reaction. We show that for energies below the Coulomb barrier ($E_{\rm cm}/V_B \leq 0.7$), the breakup cross sections are enhanced due to these couplings. However, for $E_{\rm cm}/V_B \geq 0.8$, a suppression of the breakup cross section is rather observed. It is shown that the suppression factor increases with the incident energy.

References

[1] Nunes F M and Thompson I J 1999 *Phys. Rev.* **C 59** 2652
[2] Summers N C and Nunes F M 2004 *Phys. Rev.* **C 70** 011602 (R)
[3] Canto L F, Lubian J, Gomes P R S and Hussein M S 2009 *Phys. Rev.* **C 80** 047601
[4] Diaz-Torres A and Thompson I J 2002 *Phys. Rev.* **C 65** 024606
[5] Bertulani C A 2005 *Phys. Rev. Lett.* **94** 072701
[6] Mukeru B, Lekala M L and Denikin A S 2015 *Nucl. Phys.* **A 835** 18
[7] Mukeru B, Lekala M L and Denikin A S 2015 *J. Phys. G: Nucl. Part. Phys.* **42** 015109 (11pp)
[8] Mukeru B, Lekala M L and Rampho G J 2015 *J. Phys. G: Nucl. Part. Phys.* **42** 085110 (12pp)
[9] Moro M M, Crespo R, Nunes F M and Thompson I J 2003 *Nucl. Phys.* **A 722** 455c
[10] Otomar D R, Lubian J, Gomes P R S and Correa T 2013 *J. Phys. G, Nucl. Part. Phys.* **40** 125105, 13pp
[11] Otomar D R, Gomes P R S, Lubian J, Canto L F and Hussein M S 2013 *Phys. Rev.* **C 87** 014615
[12] Iseri Y, Yahiro M and Kamimura M 1986 *Prog. Theor. Phys. Suppl.* **89** 84
[13] Thompson I J 1988 *Comput. Phys. Rep.* **7** 167
[14] Kundu A *et al.* 2016 *Phys. Rev.* **C 94** 014603
[15] Diaz-Torres A, Thompson I J and Beck C 2003 *Phys. Rev.* **C 68** 044607
[16] Mukeru B and Lekala M L 2017 *Nucl. Phys.* **A 965** 1-12